Recent Discoveries In Astronomy

Forensic Astronomy

Fred M. Johnson, PhD
Professor of Physics Emeritus

Lawyers & Judges Publishing Company, Inc.
Tucson, Arizona

Lawyers & Judges Publishing Company, Inc.
P.O. Box 30040 • Tucson, AZ 85751-0040
(520) 323-1500 • FAX (520) 323-0055
e-mail: sales@lawyersandjudges.com
www.lawyersandjudges.com

Library of Congress Cataloging-in-Publication Data

Names: Johnson, Fred M., 1926- author.
Title: Forensic astronomy / Fred M. Johnson, PhD, Professor of Physics Emeritus, Department of Physics–Astronomy, California State University–Fullerton.
Description: Tucson, AZ : Lawyers & Judges Publishing Co., [2021] Includes bibliographical references and index. | Summary: "A number of important celestial phenomena are presently misinterpreted in Astronomy. These include dark matter, dark energy, Black Holes, the Big Bang, and the origin of solar systems. This is an impressive list of misinterpretations which can all be shown to arise from either neglecting laws of physics or their incorrect usage. "Forensic Astronomy" will provide you with alternative explanations which are simple, straightforward and follow a logical order. In fact, the aim of this book is multifaceted. Here is an overview of its main goals: To educate the public on some of the more recent solar and planetary developments. To show that the universe underwent a large variety of processes, prior to the development of mankind. All developments presumably required a sequence of multiple, complex, evolutionary steps, which were well-timed and executed. Similar solar system developments are presumably in progress throughout our galaxy and very likely, throughout our visible universe, at various evolutionary stages. To discuss mankind's place in the Universe. To rectify some of the (erroneous) baffling concepts and theories that have unfortunately crept their way into astronomy. To point out the most critically neglected concepts of physics : the Second Law of Thermodynamics, the conservation of angular momentum, and the fact that all matter in the universe is not necessarily electrically neutral"—Provided by publisher.
Identifiers: LCCN 2020052660 | ISBN 9781936360840 (soft cover)
Subjects: LCSH: Astronomy.
Classification: LCC QB43.3 .J63 2021 | DDC 523—dc23
LC record available at https://lccn.loc.gov/2020052660

ISBN 13: 978-1-936360-84-0
Printed in the USA
10 9 8 7 6 5 4 3 2

Acknowledgments

Thanks and appreciations are due to Dr. Robert Tomaszewski for extensive library research.

Deborah Johnson, Emily English, Amy Gavello and Ken Tomb provided superior technical assistance.

The support and encouragements from Steve Weintraub, the Publisher, is gratefully acknowledged.

Table of Contents

Forward

Astronomy is full of unresolved enigmas and strange phenomenon. This book will attempt to resolve some of the most pressing ones. Hold on to your seats as there will be many surprises in store.

In this new edition, I will give a simple solution to the century old mystery for the real origin of gravity. Another new chapter will include a new theory for the origin of the Sun and our entire solar system. A third new item will explain details of the center of our galaxy. Lastly, there is a planetary nebulae in our solar system that accounts for the three degree pulsating radiation.

Forensic Astronomy

Introduction

A number of important celestial phenomena are presently misinterpreted in Astronomy. These include dark matter, dark energy, Black Holes, the Big Bang, and the origin of solar systems. This is an impressive list of misinterpretations which can all be shown to arise from either neglecting laws of physics or their incorrect usage.

The book that you're about to read will provide you with alternative explanations which are simple, straightforward and follow a logical order. In fact, the aim of this book is multifaceted. Here is an overview of its main goals:

1. To educate the public on some of the more recent solar and planetary developments.
   a. To show that the universe underwent a large variety of processes, prior to the development of mankind.
   b. All developments presumably required a sequence of multiple, complex, evolutionary steps, which were well-timed and executed.
   c. Similar solar system developments are presumably in progress throughout our galaxy and very likely, throughout our visible universe, at various evolutionary stages.
2. To discuss mankind's place in the Universe.
3. To rectify some of the (erroneous) baffling concepts and theories that have unfortunately crept their way into astronomy.
4. To point out the most critically neglected concepts of physics: the Second Law of Thermodynamics, the conservation of angular momentum, and the fact that all matter in the universe is not necessarily electrically neutral

Historical Perspective

Mankind has always been fascinated and simultaneously mystified by the workings of our universe. What forces are operating and what is behind and beyond our brief existence? Ancient civilizations (worldwide) were continually in search of deeper meanings and expressed their interest in various forms, including superstition, magic, and sorcery. This probably preceded the emergence of a unique group of individuals (such as priests, for example) who would pontificate and propagate their newly found wisdom. Surprisingly, some of these very ancient concepts and "beliefs" are still with us, just as we still have a small remnant DNA component (about 2% we are told) of the extinct Neanderthal species. Although survival in a harsh environment was always a primary goal, there were generally moments for reflection and speculations, which were shared by an increasing-size population of Homo sapiens, who meantime developed some rudimentary language and communication abilities.

Advanced civilizations emerged 4000-5000 years ago. There were scribes who kept records and transcribed epics and isolated historical events. These included descriptions of imagined cosmology. The Egyptians, Hindus, Mesopotamians, and Greeks provided divergent cosmic scenarios, all of which of course are interesting, but clearly mistaken (See author's text "Voyage into Astronomy Second Edition" 2015).

Thus, prior to an attempt to set the record straight with a completely new, ***physics-based*** astronomy, one must be able and willing to reject some of the flawed astronomical and geophysical science-fiction and be willing to challenge it. This mantra of critical analysis and examining realistic alternative theories is the light motif of this text.

Why, even now, is it so difficult to unravel the early origins of the Solar System and the steps leading to the evolution of life? The answer is a rather sad one. Our lives are too short - far too short. Four and a half billion years before the advent of man is just too long relative to our lifetime. In fact, even a million-year long lifespan would be an insufficient duration of time to demonstrate the extensive changes that have occurred on our planet! One thing stands out-the steps forward were very sporadic and far between. Actually, as with all progress, they were marked by one, cautious, hesitant foot forward followed by three steps backwards. Among the early Greek philosophers were, indeed, some precocious giants who had some correct ideas which, 1,500 years later, were rediscovered by others. Unfortunately, the forward march of accepted ideas does not come about on a solid front with all mankind sharing its advances equally. Rather, isolated forward thrusts were occasionally made by individuals, which took a very long time to be shared with or accepted by most of mankind. To shake us out of our present-day smugness, one has only to be reminded that there exists a *Flat Earth Society* in England and elsewhere. The concept of a flat and square Earth should surely have been eradicated long ago! The evidence for a round Earth can be easily observed during lunar eclipses when the Earth's shadow passes slowly across the moon. The implication of a round or spherically-shaped Earth associated with this relatively simple event did not make a great impact on man in ancient times and is even today largely ignored. How many people now bother to look up to watch the passage of the stars, planets, and the moon? Only a small fraction of the world's population: mostly astronomers. There were exceptional astronomers in antiquity. For example, Eratosthenes (276-194 B.C.) derived the radius of a spherical Earth to within 10 percent of its actual measured value. This was a tremendous accomplishment for its time.

Many ancient civilizations were mystified by the phenomena of the setting of the sun in the westerly direction and its re-emergence in the easterly direction. Monuments, such as Stonehenge in England, provide evidence for early attempts to track the rising and setting of the sun, and as a result established a primitive calendar. The Aztecs (and others) even resorted to ritual ceremonies involving human sacrifice in their erroneous belief to secure the continual re-emergence of the sun.

As early stargazers among ancient civilizations would be viewing the celestial sphere at night, they would notice some "stars" moving relative to the fixed stars. These are what we now recognize as the planets of our solar system. Some of the planets occasionally change apparent direction with respect to the background stars for a brief period, providing another conundrum. As long as the geocentric theory was the dominant paradigm, astronomers in the past were forced to explain these observations using erroneous "epicycles" to describe planetary motion. This phenomenon is now well-understood within the context of the heliocentric theory.

Starting with Copernicus and culminating with Galileo, Kepler and Newton, ancient astronomical dilemmas were resolved by means of science and logic and universally accepted (with the exception of a handful of holdouts, such as the previously mentioned *Flat Earth Society*).

Galileo's ideas were considered to be so controversial by the Catholic Church that his books were on their forbidden list until the early 1900's. In the year 1600, Giordano Bruno was burned at the stake in Rome for his "deviant" ideas concerning the possibility of life on other planets and his support of Galileo's heliocentric theory.

Thus, the situation that we're in today is unfortunately not too far different from that at the time of Galileo, despite the progress in science. Inertia associated with astronomical belief systems however has remained relatively constant throughout the ages. It is amazing to see hundreds of scientists climb aboard the various astronomical bandwagons, invariably unable or unwilling to question any particular paradigm (inertia). Modern astronomy requires walking in synchronism with one another: consensus and conformity is the present watchword! So, what's wrong here? What is it that has led some astronomers and geologists to go astray? The answer is, they are indeed obtaining solutions, but they are mistaken in assuming that ***any*** solution is a *unique* solution. The key word is "unique". A solution that stands alone while claiming to have solved a scientific enigma, cannot be considered a "unique" solution, because it has not withstood the challenges of alternative possibilities or a critical forensic analysis.

So, how *do*es one know whether the correct answer has been obtained? Is it by authority or edict? I don't think so. By the bandwagon effect? Certainly not. In addition to the use of logic and simplicity, allow me to suggest a significant, alternative approach: The use of a forensic technique.

Forensics

The techniques and methods employed in a forensic type of investigation are akin to a crime scene. They employ multiple approaches, different methods, known physics, engineering and technology. Problems are studied using the latest and most modern methods, but above all from multiple perspectives that *do not neglect even the smallest of details*: "Stones have to be studied from all angles." E.g., Quasars are now studied **solely** for their Doppler redshifts. Spectroscopic evidence of potential Doppler blueshifts are ignored, hidden or misinterpreted as redshifts. Thus, forensic investigations discover errors and mistakes as well as provide alternative solutions. Moreover, forensic scientists are prepared to testify in court and be cross-examined individually. They do not have the luxury of reliance on group support or consensus.

The only laws that one can safely rely on (to some extent) unless proven otherwise are the laws of physics, used wisely and correctly. Some of these physical laws have been thoroughly tested experimentally and are in fact continually being challenged. That is the approach of this text. TO CHALLENGE. This is the foundation of the forensic approach, which will be used throughout the text: To provide alternative theories for comparison and evaluation!

Finally, I believe one has to be honest enough to admit one's limitations. I can only bring it so far and have to allow others to take it further. One has to be honest enough to admit ignorance, not only on my part, but also on the part of everyone else so far. I think our ability to admit ignorance is a good beginning. It's a humbling experience to examine and attempt to interpret the universe.

Astronomy as a whole would greatly benefit if it were to pay closer attention to the building industry. All modern buildings require an extremely solid substructure, which usually involves deep-seated cement pillars. Frequent checks by government inspectors ensure compliance with these necessary preparations. Contrast these procedures vis-à-vis the present developments in astrophysics, where theories are built upon quicksand. More elaborate theories are then superimposed with incrementally increased complexities.

This text reexamines and corrects major foundation theories in astrophysics based on a combined physics and forensic analysis.

In the field of forensics, everything has to be examined and challenged. There is little room for such antics as *modeling* (a popular feature in astronomy), or wild, unproven hypotheses. If indeed a hypothesis in astronomy is rendered, it should be designated as such, with ample warnings attached. Another popular feature in today's astronomical publication is the use of the term "constraint", which is used to justify an assortment of astronomical observations, usually coupled with dubious modeling, complex equations, correlations, and extensive lists of similar publications. How did we arrive at this stage? Could it possibly be traced to our childhood fairytales?

Beliefs

> *"The difference between stupidity and genius is that genius has its limits."*
>
> —Albert Einstein

Almost everything in the universe is finite with two exceptions: ignorance and stupidity. There is no limit to either one.

Fundamental *beliefs* are extremely difficult to change (inertia). These changes are often painful, and result in serious confrontations. As mentioned above, some peculiar *belief* systems have survived in one form or another for enormous periods of time. That says a lot about human misinformation transmission and rubbish retention. Unlike modern computers, our brains are not susceptible to selective erasures. (People die, but *beliefs* are passed on, and therefore last virtually forever.)

To question any longstanding *belief* system is the equivalent of touching the proverbial "third rail".

Young children are accustomed, or perhaps programmed, to hear fairytales during their youth. These fanciful stories can be ancient, or relatively new. Adults eventually may even enjoy some of them, as for example the Harry Potter serial adventures. Mankind as a whole takes these stories in stride. Some may even eventually be elevated to a higher plateau, where they become sacrosanct and are almost untouchable. A whole industry, such as Disneyland and Knott's Berry Farm got into the act. Their financial success does indeed indicate that once children reach adulthood, some of their "sentiments" remain, to some extent, at the level of their earlier childhood years. This should not surprise anyone.

This preamble could be relevant, since it could implicate a larger scientific group of people and illustrate the human tendency to build upon our childhood mental development. Thus, even scientifically trained individuals may still manifest subconscious remnants of early mental stimulations, programs, and fairytale exposure. This could possibly (but not necessarily) affect the development of their *belief* systems.

Caveat:

In today's litigious environment, one generally finds some written disclaimers, which are meant to avoid potential legal disputes. Here is another one:

While this text does inspire critical thinking, the views it expresses are primarily for enjoyment and stimulation of the senses. They are not meant to forcibly change or mock the existing views/attitudes/religious leanings of the reader.

Let us take a look at a few examples (dilemmas) that will be addressed in this text:

A. **Earth Core Density:** The density of the Earth was at one time much greater than its present average value (5.5 g/cc). 4.25 billion years ago, its average density would have to have been in excess of 44 g/cc. There is no known material of such high density. Iron has a density of only 7.7 gm/cc. The average core density of the Earth *currently* is 13.5 g/cc, almost twice as dense as iron!

B. **Causation of Continental Drift:** It took about half a century for scientists to accept the theory of continental drift because the amount of underlying energy necessary for this to occur was unknown. It will be addressed in this text.

C. **Earth's Development:** Scientists had put forth the concept of "Pangaea", the single prehistoric continent from which all modern continents supposedly originated. This text will have an explanation for the fact that soil samples and the shape of continents actually fit together in a spherical formation, rather than an approximately flat continent. The earth was at one time much smaller than its current size. (In fact, at one time in the past, its radius was approximately half its current value, projecting a much larger average density for the Earth.)

D. **The Sun's Energy Mechanism:** In the laboratory, a self-sustaining proton-proton fusion reaction has so far never been achieved because of the impossibility of containing a finite quantity of gas at a temperature of fifteen million degrees for a finite length of time. Here, we postulate that the Sun instead operates on a nuclear fission mechanism, which also produces all atomic elements.

E. **Solar System Origin:** We know that the Sun presently has only 2% of the solar system's angular momentum, while the major outer planets Jupiter, Saturn, Uranus and Neptune together have the rest of the angular momentum. While presenting the solution to the solar system's origin, a large number of mysteries and dilemmas will be resolved. In addition, you will encounter a major surprise about our solar system and a total solution for the origin of the incoming fluctuating 3-degree radiation.

F. **Moon's Origin:** A simple and elegant solution will be presented in this text.

G. **Earth's Internal Heat Source:** The Sun chapter addresses this problem.

H. **Origin of Organic Molecules:** Part of chapter 6 provides a discussion on this topic.

I. **The DIBs:** The DIBS are a collection of over 500 specific (diffuse) spectroscopic absorption signals in the incoming light of some special designated stars. In the spaces between the stars, there exist dust clouds made up of a collection of so-called "unidentified" molecules. However, using the scientific method, careful spectroscopic study has established the identity of the clouds' major chemical constituent. The overwhelming spectroscopic evidence could be contrary to someone's *belief system*. Chapter 6 contains the historical background, while chapter 8 provides the complete solution.

J. **Shapes of Galaxies:** Photographs of galaxies reveal a large variety of shapes, including some with spiral arms. This text provides a scientific explanation.

K. **Black Holes:** Astronomers postulate that a black hole is situated at each galaxy's center. However, recent data (interpreted in this book) clearly proves their nonexistence.

L. **Quasars:** You will discover new solutions in a chapter of this book. This fascinating topic will also be reinterpreted in view of Halton Arp 1987 book entitled *Quasars, Redshifts, and Controversies*.

M. **Supernovas:** This book will discuss new theories.

N. **The origin of gravity:** This discovery will come as a total surprise in Chapter 7.

These topics will be addressed, alternative solutions will be presented. The point that will be made is that in order to understand the huge cosmological picture, one first must have a solid understanding of our own solar system and its origin before venturing out to extreme distances within our galaxy and beyond, into outer space. **Good science proceeds from the known to the unknown and not vice-versa!**

What you are about to read may surprise many, it is, after all, a completely new way of examining the early development of our solar system and the universe. Of course, one could just as easily not be concerned at all with this type of knowledge. Do people as a whole really care how the solar system developed? Does it make a difference to one's standard of living? Absolutely not. Yet, there surely is a nagging thought by most thinking people: How and what really happened in the ancient past?

All that is required of the reader is an open mind (some familiarity with algebra could also be helpful). Following this lengthy (but essential) prologue and introduction, you are primed and should finally be ready to proceed. Here goes.

Chapter 1

Planetary Core Sizes, their Energy Flux Relationship, and Its Implications

Fred M. Johnson

Abstract

Measured heat flow data from each planet in our solar system plus over 50 stars, including our Sun, was plotted against each object's known mass to generate a continuous exponential curve at an R-squared value of 0.99. The unexpected yet undeniable implication of this study is that all planets and celestial objects have a similar mode of energy production. It is widely accepted that proton-proton reactions require hydrogen gas at temperatures of about 15 million degrees, neither of which can plausibly exist inside a terrestrial planet. Hence, this paper proposes a nuclear fission mechanism for all luminous celestial objects, and uses this mechanism to further suggest a developmental narrative for all celestial bodies, including our Sun. This narrative was deduced from an exponential curve drawn adjacent to the first and passing through the Earth's solid core (as a known prototype). This trend line was used to predict the core masses for each planet as a function of its luminosity.

Keywords: planetary cores; planetary heat flux; nuclear fission reactor; hydrocarbons; Venus

I. Introduction

The Earth's geothermal heat flux and its origin have been of interest to geologists, geophysicists, as well as a few astronomers for well over half a century. The customary explanation for this phenomenon generally involves radiogenic heat production from radioisotopes of Uranium, Thorium, and Potassium.

The earliest heat flow measurements were taken on the Earth's surface. Planetary heat flow measurements were discussed in great detail by Axel Hagermann (other studies on this topic can be found in the reference section of Hagermann's paper) [1]. The Earth is by far the most thoroughly studied, with a mean value of 65 milliwatts per square meter (mWm^{-2}) passing through the continents and a mean value of 101 mWm^{-2} flowing through the oceans [2]. Since NASA's space probe launchings, and moon landings by astronauts, heat flow measurements have been obtained for most solar system planets, as well as the Moon. Two separate heat flow measurements were made by the astronauts of Apollo 15 and 17 on different locations on the moon's surface. The lunar heat flow values were 21 mWm^{-2} and 16 mWm^{-2} respectively. The latter value was subsequently adjusted by Waren and Rasmussen [3] to 12 mWm^{-2}. Jupiter's closest satellite Io exhibits an unusually large heat flow of 2 Wm^{-2} [4]. The Martian heat flow data was determined to be 6.4 +/- 0.4 mWm^{-2} by Hahn et al. [5]. The Sun's known parameters were included not only for the sake of completeness, but the Sun after all is the principle component of our solar system. With the exception of Mercury and Venus, there is heat flow data for all the other planets [6]. By integrating the heat flow measurements over the whole surface of the planet, one can establish its luminosity.

II. Methods

A luminosity (L) value for each planetary object was obtained by multiplying its surface area (A) by its average measured heat flow value (H):

$$L = (H) \cdot (A) \qquad 1$$

This assumes that the heat flux measurements taken represent the average value over the entire surface of the planet. Table 1 is a listing of the calculated luminosities of each planet, along with some other properties.

Table 1
Planetary Properties

	Radius (m)	Mean Density (gm/cc)	Mass (kg)	Luminosity (Watts)
Earth	6.378×10^{6}	5.5	5.970×10^{24}	3.311×10^{13}
Mars	3.397×10^{6}	3.94	6.419×10^{23}	9.333×10^{11}
Jupiter	7.150×10^{7}	1.24	1.898×10^{27}	3.4673×10^{17}
Saturn	6.030×10^{7}	0.62	5.685×10^{26}	9.1203×10^{16}
Uranus	2.556×10^{7}	1.3	8.685×10^{25}	3.388×10^{14}
Neptune	2.476×10^{7}	1.61	1.024×10^{26}	3.311×10^{15}

Since stars are characterized by a mass-luminosity relationship, it was hypothesized that there could be a similar relationship among the planets. After plotting the masses versus the luminosities of each planet on a logarithmic scale, it became clear that indeed such a relationship exists.

Next, because the mass of the planet Jupiter is not far removed from that of some nearby stars, a study of the mass-luminosity relationship for those stars was of interest.

Table 2 is a listing of 39 nearby G, K, and M stars, which provides the luminosity and mass of each stellar object. [7] The G, K, M stellar classification is based on the stellar surface temperature in decreasing order of temperature, the Sun being a G-type star. For completeness, 12 highly luminous stars were added later, representing larger, younger, and hotter stars than the Sun. Data therefore includes a wide range of all stars in our galaxy.

Stellar luminosity (L) was calculated with the solar luminosity ($L_\odot = 3.846\times10^{26}$) as a reference using the following equation:

$$\frac{L}{L_\odot} = 10^{\frac{2}{5}(M_{v\odot}-M_v)} \qquad \text{2}$$

Where:

$M_{v\odot}$ and M_v are the absolute visual magnitudes of the Sun and the target star, respectively.

The mass-luminosity relationship of a star, including our Sun, is subject to the following equations:

$$\frac{L}{L_\odot} = .23\left(\frac{M}{M_\odot}\right)^{2.3} \text{ when } (M < .43M_\odot) \qquad \text{3}$$

$$\frac{L}{L_\odot} = \left(\frac{M}{M_\odot}\right)^{4} \text{ when } (.43M_\odot < M < 20M_\odot) \qquad \text{4}$$

Where:

$M_\odot$ = mass of Sun $\quad$ M = mass of star

$L_\odot$ = luminosity of Sun $\quad$ L = luminosity of star

After preliminary results confirmed a mass-luminosity relationship among the planets using data from Table 1, the data from Table 2 was added to the plot, creating Curve A of Figure 1. Since the Earth's core has a known mass, and is clearly the origin of its energy production, it was used as a benchmark for an analogous curve, which has a similar slope to that of Curve A, and terminates in the region of the 12 most luminous stars. This is designated as Curve B.

III. Results

Figure 1 is unique in that it links the planets and the stars on a single, **continuous** mass-luminosity curve (A). Curve A is represented by the following equation:

$$y = 0.6318e^{0.1236x}$$

With a data to regression line fit of $R^2 = 0.9871$, indicating a very small variance.

Where $y = Log(L)$ and $x = Log(M)$

This unique curve is represented by a single equation covering an extremely wide range of masses (11 orders of magnitude) and luminosities (33 orders of magnitude). This implies an extremely stringently-fitting function, which covers both stars and planets. An equation that covers such a wide range of variables most likely demonstrates intrinsic significance.

The space in between curves A and B is representative of the non-core component (inactive masses) of each planet, as well as the Sun (two examples of this are indicated in Fig. 1 by double-sided arrows). The core mass for each of the planets can therefore be predicted by subtracting the horizontal difference in distance between Curve A and Curve B from the total mass of each planet.

Table 2
Stellar Properties

Name	Spectral Type	Absolute Magnitude	L (Watts)	Log of L	Mass (Kg)	Log of Mass
Sun	G2.0VN	4.85	3.846E+26	26.59		
39 G, K, and M Stars (within 100 parsecs)						
GJ 551	M5.0 V *	15.48	2.153E+22	22.33	5.339E+28	28.73
GJ 559 A	G2.0 V G	4.380	5.929E+26	26.77	2.216E+30	30.35
GJ 559 B	K0 V N	5.710	1.742E+26	26.24	2.670E+30	30.43
GJ 699	M3.5 V *	13.25	1.679E+23	23.23	1.304E+29	29.12
GJ 406	M5.5 V *	16.64	7.396E+21	21.87	3.355E+28	28.53
GJ 411	M2.0 V *	10.44	2.234E+24	24.35	5.491E+29	29.74
GJ 65 B	M6.0 V *	15.93	1.422E+22	22.15	4.458E+28	28.65
GJ 729	M3.5 V *	13.08	1.963E+23	23.29	1.396E+29	29.14
GJ 144	K2.0 V G	6.200	1.109E+26	26.05	1.458E+30	30.16
GJ 887	M1.0 V *	9.760	4.178E+24	24.62	6.421E+29	29.81
GJ 447	M4.0 V *	13.53	1.297E+23	23.11	1.166E+29	29.07
GJ 820 A	K5.0 V *	7.480	3.412E+25	25.53	1.314E+30	30.12
GJ 725 A	M3.0 V *	11.17	1.140E+24	24.06	2.999E+29	29.48
GJ 725 B	M3.5 V *	11.96	5.508E+23	23.74	2.186E+29	29.34
GJ 15 A	M1.5 V *	10.31	2.518E+24	24.40	5.658E+29	29.75
GJ 15 B	M3.5 V *	13.30	1.603E+23	23.21	1.278E+29	29.11
GJ 1111	M6.0 V *	17.10	4.842E+21	21.69	2.791E+28	28.45
GJ 71	G8.5 V G	5.680	1.791E+26	26.25	1.643E+30	30.22
GJ 1061	M5.0 V *	15.26	2.636E+22	22.42	5.831E+28	28.77
SCR 1845-6357 A	M8.5 V *	19.47	5.458E+20	20.74	1.080E+28	28.03
SO 0253+1652	M6.5 V *	17.21	4.375E+21	21.64	2.670E+28	28.43
DEN 1048-3956	M8.5 V *	19.37	5.984E+20	20.78	1.124E+28	28.05
GJ 1245 C	M V -	18.47	1.371E+21	21.14	1.612E+28	28.21
GJ 1245 B	M6.0 V *	15.72	1.726E+22	22.24	4.850E+28	28.69
DEN 0255-4700	L7.5 V *	24.44	5.611E+18	18.75	1.476E+27	27.17
GJ 166 B	DA4 N	11.03	1.297E+24	24.11	4.793E+29	29.68
GJ 166 C	M4.0 V *	12.75	2.661E+23	23.43	1.593E+29	29.20

Continued

Table 2 (continued)
Stellar Properties

Name	Spectral Type	Absolute Magnitude	L (Watts)	Log of L	Mass (Kg)	Log of Mass
G 099-049	M3.5 V *	12.71	2.761E+23	23.44	1.619E+29	29.21
LHS 1723	M4.0 V *	13.59	1.227E+23	23.09	1.138E+29	29.06
GJ 445	M3.5 V *	12.15	4.624E+23	23.67	2.026E+29	29.31
GJ 251	M3.0 V *	11.27	1.040E+24	24.02	2.882E+29	29.46
2MA 1835+3259	M8.5 V	19.50	5.309E+20	20.73	1.067E+28	28.03
GJ 752 B	M8.0 V *	18.61	1.205E+21	21.08	1.524E+28	28.18
GJ 139	G8.0 V G	5.390	2.339E+26	26.37	1.756E+30	30.24
GJ 780	G8.0 IV G	4.620	4.753E+26	26.68	2.097E+30	30.32
GJ 661 B	M V -	11.32	9.931E+23	24.00	2.824E+29	29.45
LHS 3003	M7.0 V *	17.97	2.173E+21	21.34	1.970E+28	28.29
G 041-014 B	M V -	12.99	2.133E+23	23.33	1.447E+29	29.16
LP 771-095 B	M3.5 VJ *	12.58	3.112E+23	23.49	1.705E+29	29.23
GJ 2130 B	M2 VJ N	12.79	2.565E+23	23.41	1.568E+29	29.20
GJ 2130 C	M V -	13.79	1.021E+23	23.01	1.050E+29	29.02
12 Highly Luminous Stars						
Alpha Centauri A		4.380	5.929E+26	26.77	2.188E+30	30.34
Sirius A		1.400	9.226E+27	27.97	4.018E+30	30.60
Vega		0.580	1.963E+28	28.29	4.246E+30	30.63
Capella Aa		0.400	2.317E+28	28.37	5.110E+30	30.71
Arcturus		-0.310	4.457E+28	28.65	2.188E+30	30.34
Aldebaran		-0.630	5.984E+28	28.78	3.381E+30	30.53
Polaris Aa		-3.600	9.226E+29	29.97	1.074E+31	31.03
Bellatrix		-2.640	3.811E+29	29.58	1.671E+31	31.22
Canopus		-5.530	5.458E+30	30.74	1.949E+31	31.29
Antares		-7.200	2.541E+31	31.41	2.466E+31	31.39
Rigel		-7.840	4.582E+31	31.66	3.580E+31	31.55
Betelgeuse		-8.000	5.309E+31	31.73	1.531E+31	31.18

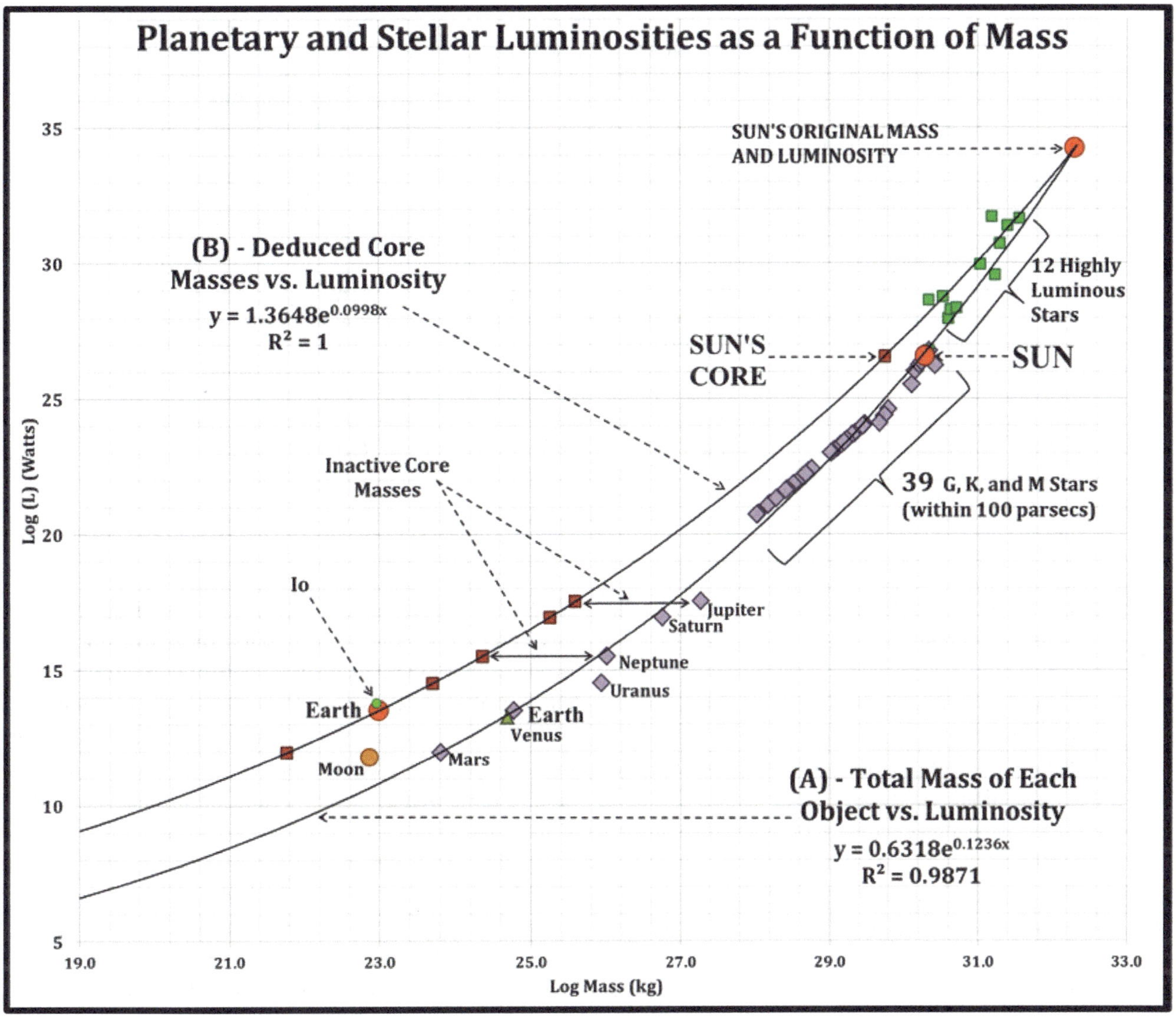

Figure 1

Based on this analysis, Table 3 lists both the predicted core mass and the deduced radii for each object, assuming a value of 13,000 kg/m^{-3} for each of their densities. The density values may have to be adjusted upwards for the heavier gas planets. Almost certainly, the Sun's core mass density must be far greater than those of the planets, since its core radius is expected to be significantly smaller.

IV. Discussion

4.1 Figure 1

Figure 1 is useful for discerning the evolution of each of the planets, in particular, the so-called outliers. The following is a brief overview covering a few selected objects:

Io is unique by virtue of its relatively large luminosity. One may assume that it is of relatively recent vintage, by virtue of its size, mass, and thin crust. Consequently, it appears along the active core trend line, and is likely representative of the "early phase" of each planetary object.

The Moon is of particular interest, since its location within Fig. 1 falls roughly between the two exponential curves. This implies that the Moon did once have an active core, which, after about a couple billion years, ceased to produce energy. The Moon, therefore, did have an opportunity to undergo some geological development, but its activity ceased some time ago.

Venus is our sister planet. It developed for billions of years just like the Earth - radar photography confirmed a

Table 3
Planetary and Solar Core Masses and Radii

	Deduced Core Masses Now (kg)	Deduced Core Radii (m) (Assuming Density = 13 g/cc)
Sun	5.637×10^{29}	2.179×10^{8}
Earth	9.910×10^{22}	1.22×10^{6}
Mars	5.721×10^{21}	4.719×10^{5}
Jupiter	3.854×10^{25}	8.912×10^{6}
Saturn	1.774×10^{25}	6.881×10^{6}
Uranus	5.006×10^{23}	2.095×10^{6}
Neptune	2.291×10^{24}	3.478×10^{6}

complicated terrain on Venus. However, here are some differences between Venus and the Earth:

a. Venus does not have a measurable magnetic field.
b. It almost stopped its rotation completely. A day on Venus comprises 243 Earth days, and in the opposite direction.
c. It clearly displays an incredibly hot atmosphere, consisting of 98% carbon dioxide.
d. There are no internal heat flux measurements for Venus, hence, there is no information about its core.

It seems that a severe catastrophe must have overwhelmed Venus.

The planet **Mars** is in line to follow the Moon's example of complete core energy extinction in the not-too-distant future. It should be possible to verify the core size prediction given in this paper by means of seismic probes to be used on Mars's surface in a number of different locations. The expected results would verify predictions deduced from Fig. 1.

The planet **Saturn** probably had a somewhat similar disaster to that of Venus. However, its causation was most likely different from Venus's, since it still displays evidence of a significant magnetic field component.

4.2 Energy Production Mechanism

The most significant aspect of Figures 1A and 1B is the continuity and tight fit of all the planets and the stars to a single exponential curve. This implies that the internal energy mechanism for all planetary and stellar objects is one and the same. The horns of the dilemma are as follows: Planets allegedly produce their energy by radiogenic heating, which is clearly impossible for the stars. Moreover, the energy mechanism for the stars is supposedly proton-proton nuclear reaction at a temperature of 15 million degrees, which again would be impossible for any of the planets. This is a grave dilemma. There is a possible solution, however. Nuclear fission is the primary energy production mechanism in both planets and stars.

Using our Earth as a prototype, it is apparent that the Earth's solid core is the most logical location for its energy origin. Consequently, each planet, as well as the Sun, must contain an active, energy-emitting core. Figure 1B is an exponential curve that traces the predicted **active** core masses of each object using the measured Earth's core mass as a starting point.

Extensive seismological measurements, taken on the Earth's surface for about half a century, provided our present knowledge of the Earth's interior [8, 9]. The Earth has a solid inner core, whose outer radius extends to 1,221 km, and a so-called liquid core region which extends from 1,221 to a radius of 3,480 km The reason for its liquid (or high-pressure gas) designation arises from the fact that only seismic pressure waves can be transmitted through it, since sheer waves do not propagate through a non-solid substance.

The average density of the earth's solid core is about 13 gm/cc [9, 10]. The density of pure uranium is 18.7 and that of carbon and similar compounds is about 2.5. If a mixture of these compounds (and possibly other low-density atoms such as sulfur) constitute the bulk of the core, their relative proportional values are approximately given by the following calculation:

$$pU + pC = 100 \quad (5)$$

$$\frac{pU\rho_U + pC\rho_C}{100} = \rho_{Core} \quad (6)$$

$$\text{Thus, } pU \times 18.7 + (1 - pC) \times 2.5 = 13.0\ gm/cc \quad (7)$$

Where:

pU, pC = the percentage of Uranium and light elements like Carbon respectively

ρ_U, ρ_C, and ρ_{core} = known densities of Uranium, light elements like Carbon, and the Earth's core respectively

This calculation shows that Uranium makes up 64% of the Earth's solid core, and light elements like Carbon, 36%. This determines that the Earth's core has a Uranium content consistent with nuclear fission theory.

The evidence for nuclear fission processes in the Earth's core interior arises from a number of independent measurements. First, neutrino detectors in Japan and Italy [11, 12] determined that the bulk of the Earth's heat flow arises from nuclear fission processes from the Earth's core interior.

Second, the abundances of trace elements in the Earth's continental crust [9] are comparable to the ratios of element abundances produced by spent nuclear fuel reactors. These trace elements include Rubidium, Barium, Strontium, and Zirconium.

The third and most important determinant for establishing Earth's core fission processes are specific trace elements whose half-lives are relatively short, and thus could not have been present on the earth since its inception over 4.0 billion years ago. Examples are given below [13]:

A. Selenium-79 isotope has a half-life of 327,000 years.
B. The nuclear fission process produces a significant amount of Zirconium with isotopic values ranging from 90 to 96. Zirconium 95 and 97 decay to Molybdenum. However, Zirconium-93 has a half-life of 1.53 million years. This isotope is one of seven major long-lived fission products. Their presence on Earth is a distinct, unambiguous marker for its fission origin. (That is, they could not have been produced by any one of the many fanciful theories, such as a hypothetical supernova explosion, since the age of the earth is at least 4 billion years.)
C. Other examples of yields from fission reactions include:Technetium-99 (half-life of 211,000 years) is a 6% yield per nuclear fission, Ruthenium-(101-106), Rhodium-103 (which has a 3.6% yield), Palladium-(105-110), Tin-126 (half-life of 230,000 years).
D. The ubiquitous presence of all the rare gases: Helium, Neon, Argon, Krypton, Xenon and Radon.
E. The apparent overabundance of hydrogen and helium is a direct consequence of core evolution development: ejected neutrons morphed into protons, electrons, and neutrinos, producing hydrogen (beta decay). Helium, on the other hand, was a direct result of alpha decay within each star. Our scarce helium gas supply, generally only found in conjunction with specific hydrocarbon deposits (oil), thus originated from our Earth's core fission reactor via alpha particle production.

There is additional evidence for universal fission processes in our solar system. Note, for instance, the similarity of atomic element abundances as observed in meteorites and on the surface of the Sun [9, 14]. It is also worthwhile examining Allen for a listing of the gas components, including rare elements, on virtually all of the Solar planets, as well as the satellite Titan [9].

Having shown that nuclear fission is the most likely internal heat production mechanism within the Earth and, by extension, within the planets, the energy mechanism within the stars must now be addressed, in view of the single exponential curve connection between the two. It is generally regarded that Jupiter displays similarities to stars, despite its lower surface temperature. In fact, its mass is merely one order of magnitude less than one of the 39 plotted nearby stars, making it the perfect planet-to-star transition object (See Fig. 1).

V. Conclusion

The most significant finding of this study is the continuity and tight fit of all the planets and stars to a single exponential curve. This implies that the internal energy mechanism for all planetary and stellar objects is one and the same (nuclear fission).

Appendix

The Sun and its Development

Present total mass = $1.989 \times 10^{30}\ kg$

Detailed calculation of the Sun's fission energy utilization:

a. 200 Mev = 2×10^8 ev per fission event
b. $L_0 = 4 \times 10^{26}$ Joules/sec = 2.5×10^{45} ev/s
c. Number of fission events per second =

$$\frac{2.5 \cdot 10^{45}}{2 \cdot 10^8} = 1.25 \cdot 10^{37}$$

Here are some of the details deduced from Fig. 1:

Active core mass of the Sun $= 5.63 \times 10^{29}\ kg$

Sun's Inactive (or non-energy-producing) component $= 1.426 \times 10^{30}\ kg$

(This value includes all of the gases and all of the spent, vaporized fuel, which yields its average density as 1.4 gm/cc)

Assuming that the Sun's present luminosity stayed constant at a value of 3.845×10^{26} *Watts* for a period of 4.5 billion years, it would have required a quantity of uranium fission amounting to $5.90 \times 10^{29}\ kg$, assuming that each fission event produces 200 MeV plus an additional 25 MeV from secondary processes.

Thus, the overall net efficiency of Solar core mass use for its energy production is equal to

$$\left(\frac{5.90 \times 10^{29}}{14.26 \times 10^{29}}\right) \times 100\% = 41\%$$

One can use this information to deduce the projected Solar lifetime:

The quantity of usable fission material available from the Sun's core is equal to

$$41\% \times \left(5.63 \times 10^{29}\right)\ kg = 2.31 \times 10^{29}\ kg$$

The amount of time required for complete exhaustion of fissionable material using past Solar experience for comparison is equal to:

$$\left(\frac{2.31 \times 10^{29}}{5.90 \times 10^{29}}\right) \times 4.5\ billion\ years = \mathbf{1.76}\ \boldsymbol{billion\ years}$$

This value is derived from extrapolated data. It provides the expected life left for the Sun to use up its total available fissionable material, based on past experience. Hence, there is nothing to worry about from this source for the time being. However, Venus's experience is a far scarier phenomenon for us to consider.
Conflicts of interest: none.

Acknowledgment

ELSEVIER Publishers for permission to reproduce (as Chapter 1) the article by this author in Planetary and Space Science, Vol. 151 (2018), 125–129.

References

1. Hagerman, A. *Philosophical Transactions: Mathematical, Physical and Engineering Sciences* **363**, 2777 (2005).
2. Pollack, H., Hurter, S., Johnson, J. *Reviews of Geophysics* **31,** 267 (1993).
3. Warren, P., Rasmussen, K., *Journal of Geophysical Research: Solid Earth* **92**, 3453 (1987).
4. Morrison, D., Telesco, C. *Icarus* **44**, 226 (1980).
5. Hahn, B., McLennan, S., Klein, E. *Geophysical Research Letters* **38** (2011).
6. Spohn, T. *Encyclopedia of Astrobiology* (2014).
7. "Recons Census of Objects Nearer than 100 Parsecs". *www.recons.org* (2012).
8. Dziewonski, A.M. and Anderson, D.L. *Phys. Earth Planet Int.* **25**, 297 (1981).
9. Hill, F. *Allen's Astrophysical Quantities*. 4th Ed. Springer. pp. 29–30, 256, 342 (1999).
10. Buchbindler, G. *Physics of the Earth and Planetary Interiors* **5**, 123 (1972).
11. Biello, D. "Nuclear Fission Confirmed as Source of More than Half of Earth's Heat". *Scientific American*. Nature America (2011).
12. KamLAND Collaboration. *Nature Geoscience* **4**, 647 (2011).
13. "Fission products (by element)". *Wikipedia* (2017).
14. "Abundance of the chemical elements". *Wikipedia* (2017).

Chapter 2

A Modern Theory of the Sun

Based on the recent discovery of nuclear fission in our Sun and in all the planets, plus an accumulation of solar-related enigmas, a modern theory of our Sun required development and produce solutions.

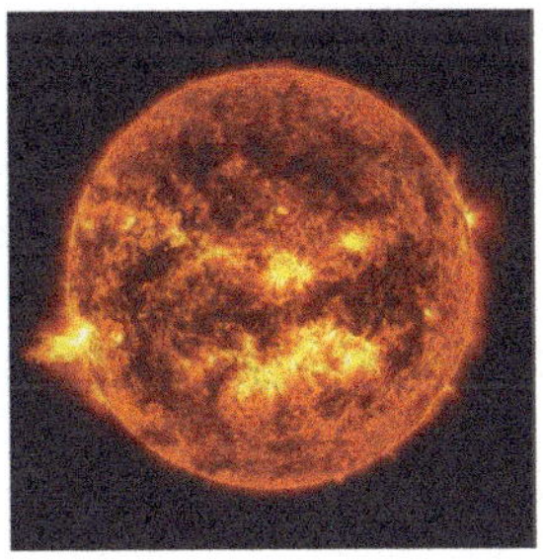

Credit: NASA, SDO (Solar Dynamics Observatory), February 10, 2015

Synopsis

Solar Enigmas

Consider the following inexplicable solar phenomena and enigmas:

- For over half a century, the million-degree temperature of the Sun's Corona had been one of the many challenges to the standard solar model. Other quandaries included the vast temperature difference between a sunspot (4500K) and the Sun's surface temperature of 5800K, the 11 year sunspot cycle and many others:
- If the Sun is considered to be a giant gas cloud, why are there these active, multi-million complex cellular structures, (GRANULES) covering its entire surface?
- What is the origin and purpose of the recurrent appearance of Sunspots?

How are the giant flares generated and what constituents the so-called wind?

- How are the immense surface magnetic fields on the Sun generated?
- Considering the Sun as a closed system, how does one reconcile the alleged standard model proton fusion theory with the Second Law of Thermodynamics?
- The standard solar theory cannot account for the long term constant solar Luminosity (its output) over billions of years. (This theory is entirely missing).
- Are you aware that for over 70 years, scientists and engineers have failed to produce controlled, self-sustaining, nuclear proton-proton fusion, despite an infusion of multi-billion dollars?
- If sustained nuclear fusion fails to operate in a complex, controlled lab environment, why would one expect it to work in a completely **uncontrolled** gaseous environment?

Introduction

[Both the Sun and the center of our galaxy (Sagittarius, SgrA*) have features in common. Thus, by examining the Sun first, their similarity will aid in providing solutions to the many enigmas associated with SgrA*].

A modern study of our Sun requires casting a relatively wide net.

There now exist large amounts of accumulated solar observations, which include data from NASA's solar satellites Ulysses, SOHO, ACE, Trace, Cluster, SOLO, ESA/NASA's Solar orbiter, joint JAXA/NASA Hinode and the most recent 2018 Parker Solar Probe. Parker's theories [1] were unable to account for many solar enigmas: These include the multimillion speckled, active, granular surface, unexplained violent local events such as dramatic CMEs (coronal mass ejections), origins of the large number of magnetic field lines, including their alternating field directions, the solar wind's supersonic transition at a distance of about a half solar radius from the Sun. Parker's theories [2] were also unable to account for the presence of both a slow and fast solar wind, their non-thermal acceleration, as well as plasma flow tangential to and in the same sense as solar rotation.

A recent publication [3] exhibits spectacular detailed solar wind and composite surface structures from eleven eclipses (2006–2020) as a function of solar cycle, white light, and spectroscopy. These correspond to ionized states of iron atoms, such as such as Fe IX (Fe^{8+}), Fe XI (Fe^{10+}), Fe XIII (Fe^{12+}) and Fe XIV (Fe^{13+}). The spectral interpretation implies coronal temperatures ranging from 1.2 to 1.8 MK. All eclipse photos display an enormous number of individual streamers emanating from the Sun, including detailed locations of specific dominant coronal forbidden spectral lines and their corresponding charge states. Several photos show prominences at the base of the corona which are connected to the filamentary structures in the overlying corona. Solar wind streamers were recorded with speeds between 300 and 700 km/sec originating at the Sun's surface.

The new modern solar theory will provide solutions to many of the current solar enigmas. The impetus for this study was the key discovery of a Mass vs Luminosity plot linking the Sun and all solar system planets in a tight-fitting curve [4]. It implies a similar energy production mode (nuclear fission) for the Sun, all the planets, and all the stars in our galaxy. In fact, the slope of the curve indicates that the luminosity of these objects is proportional to square of their masses, which is consistent with the concept of nuclear fission. The curve includes 39 K, G and M stars within a distance of 100 parsecs. All these stars, including our Sun display low surface rotational speeds, about 2 km/s). Whereas Oe, Be stars display speeds of 500 km/s [5]. Another possible relevant fact is that all early type, hot stars have a mass luminosity relationship such that L is proportionate to $M^{3.5}$(rather than M^2). This suggests the possibility that young, hot stars constitute a complex, multiple set of triple binary systems. According to Kepler's third law for a single binary system, the following equation applies:

$$\frac{P^2}{a^3} = \frac{4\pi^2}{G\left(M_1 + M_2\right)}$$

Thus, $(M_1 + M_2)^{1/2} = 2\pi a^{3/2}/PG^{1/2}$ *where* P is the period of the binary system, and a is the mean orbital distance. Rewriting the mass luminosity relationship as: L is proportional to $M^{3.5} = M^2 \cdot M^{3/2}$, where the triplet set of binaries are $(M_1 + M_2)^{1/2} \cdot (M_3 + M_4)^{1/2} \cdot (M_5 + M_6)^{1/2}$ and where each individual system's respective mean distances and periods are given according to Kepler's third law. Hence, the dominant energy mechanism for hot, young stars is also nuclear fission, however enhanced by a complex set of triple binaries.

The origin and method by which our Sun produces its prodigious power output of about 3.845×10^{26} Watts has been of interest for centuries. Some of the early erroneous speculations included fossil fuel combustion and gravitational contraction. However, by 1942, the discovery of nuclear energy processes produced new possibilities.

Two different theoretical nuclear reactions were being considered for the Sun in the post-1942 era: the proton-proton fusion, and the more complex CNO (carbon-nitrogen-oxygen) reaction cycle, proposed by Hans Bethe [6]. The proven nuclear Uranium fission reactor mechanism, which under Enrico Fermi's direction operated as early as 1942, was prematurely rejected by Lloyd Motz [5] (p.384). However, reconsidered recently [4] **as the only energy mechanism capable of operating in both planets and stars.**

1. Solar Nuclear Fission Theory

Flaws and deficiencies in the old solar model [7] had, in any event, necessitated a complete re-evaluation, in view of the recent, surprising discovery [4] that nuclear fission rather than nuclear fusion is the Sun's energy production mechanism. Furthermore, the standard theory fails to account for the constant solar output during the previous 4.5 billion years. This critical theory is entirely missing from the old solar model.

The proposed solar nuclear fission theory [SNFT], comprises a dense, active, spinning, negatively charged, nuclear, (largely neutron) solar core, which periodically launches electrically charged, spinning off-springs: "Sunspots."

Sunspots travel to the Sun's outer surface, (via the inner solar plasma region) eliminate their excess charge while assembling at latitudes either above or below the Sun's equator, in paired alternating magnetic field directions. Their apparent lower surface temperature results from evaporation cooling, primarily the ejection of high velocity neutrons. Sunspots eventually submerge below the sun's surface and break up into smaller spinning nuclear fission reactors (SNFR).

Since adjacent SNFRs have opposing spins, they are magnetically attracted but charge wise repelled, thus their equilibrium is established, and their separation accounted for on the Sun's subsurface.

According to Richard Muller [8] there are about 5 million granules with average radius of 500 km on the Sun's surface, separated by about 700 km.

According to the new SNFT, there is an SNFR below each granule, generating the diverse, highly active, visible, granular phenomena. The Sun's total energy output (its luminosity L_0) is thus generated by these multimillion, active, rotating nuclear fission reactors, which heat its enveloping, neutron moderating gas to 5,800 K. The total mass (M_q) of all SNFRs is $(3.3)10^{23}$ kg. (Deduced by an analysis in the Appendix.)

Magnetic fields are generated as a result of rotating electric charges (approximately 10^{10} C each) associated with each spinning nuclear fission reactor (SNFR). Based on Muller's [8] estimated 5 million granules on the Sun's surface and its assumed equal number of subsurface SNFRs, their approximate average diameter would be in the 300 kilometer range, with a mass of about 6.6×10^{16} kg each. Their low computed average density implies that these nuclear fission reactors are hollow structures, while spinning at a rate of about 5 seconds per revolution. Neutrons on the SNFR surface escape with velocities in the 300–500 km/s range, consistent with the SNFR spin rate period. Since a neutron's half-life is about 12 minutes, it will eventually decay into a proton, electron, and an anti-neutrino. This accounts for the multi-thousands of individual proton streamers observed during solar eclipses [3]. The spinning SNFRs are assumed to be approximately spherical with an average surface thickness of about 1km, assuming nuclear surface density of about 20 gm/cc. SNFR's interior gas would be mainly hydrogen and helium.

Neutrons are emitted from the surface of each spinning SNFR, with speeds depending on their surface location. Other slow or prompt neutrons originate directly from each nuclear reactor (based on our knowledge of nuclear reactor technology). Ejected neutrons transform into protons, electrons and neutrinos. They constitute the widely observed **solar wind**, providing a solution to a longstanding enigma.

This would explain the origin of the slow and fast solar wind whose total outbound flux is estimated at 1.3×10^{36} particles per second. An equal number of neutrons are also emitted inwards from each spinning SNFR. The buildup of these neutrons which likewise decay into protons, electrons etc. increase the internal pressure P of each SNFR, which in turn obeys the well-known thermodynamic relationship: PV=nRT. The relatively thin wall thickness of each SNFR is occasionally breached, resulting in the observed, bright solar flares. Relatively thicker SNFR surfaces produce greater internal pressure buildup, which, once breached, result in the spectacular CMEs and brilliant flares. The Carrington flare of August 1860 is remembered as an extremely spectacular solar flare and its subsequent effects on the Earth's atmosphere. Based on solar observations and the CME's measured travel time, the solar wind velocity (in 1860) was approximately 2380 km/sec, consistent with an assumed relatively high internal SNFR pressure.

Parker's [2] unknown, enigmatic solar wind *acceleration mechanism* is readily explicable based on neutrons' half-life being about twelve minutes. Given a neutron wind speed of 500 km/sec, its beta emission process would occur after a time of $12 \times 60 \times 500$ seconds, which corresponds to a distance of 360,000 km from the Sun, (about half of a solar radius), assuming that some of its proton emission is in the direction of neutron's travel.

Furthermore, anti- neutrino production from a fission reactor is about 1/3 that of a comparable fusion reactor. This result is consistent with the observed neutrino flux recorded by Raymond Davis Jr. (1968), who, over an extended number of decades, consistently found only 1/3 of the neutrinos that the standard theory predicted. Each fission reaction is about NINE times more efficient than hypothetical proton-proton nuclear fusion.

The proposed SNFT provides explanations for many other enigmas such as the previously inexplicable Sun's surface and sub-surface phenomena, whose photographs strongly suggest: Direct, subsurface heating, cellular substructure generation and development. (See typical photographs of a highly active Sun, particularly those taken during total solar eclipses [3].

The Sun's central, spinning core also produces a magnetic field, which reverses direction prior to and during each Sunspot-ejection.

II. The Sun's Core Development

A provisional appraisal of the Sun's neutron core development scenario assumes that it has a rapid spin rate, a radius of about 2×10^8 m, and a mass of about 5.637×10^{29} kg [4]. The preponderance of neutrons, on its outer core surface, would favor the production of heavy atoms, such as Uranium. The core's surface is composed of a relatively thin nuclear fission crust, which produces a small amount of energy. Due to the core's rapid rotation, neutrons are preferentially ejected from its equatorial surface region. This results in a sizable depression in the central outer nuclear core surface. As a result, the outer crystal core surface splits into two spherical hollow segments. Each spinning hemisphere would sequentially separate and depart from the core, primarily, as a result of electric charge repulsion and kinetic rotational energy. The upper shell eventually reaches the Sun's photosphere and appears at the mid-latitude of the Sun's surface with its spin magnetic vector pointing up, while the lower segment would appear as a sunspot on the lower latitude part of the Sun's surface, with its spin vector pointing down. Meanwhile, the Sun's spinning core has also undergone a 180-degree flip initiated during the destabilizing action by its primary hemisphere ejection process which was initiated first as a processional motion and ended up finally as a 180 flip. A repetition of the total launch process results in the assembly of SNFRs (sunspots) on the Sun's surface, with alternating spin directions, below and above the Sun's equatorial region.

III. Solar Surface Oscillations

The Sun's surface oscillations have been amply studied since their discovery (Leighton et al.1962, ApJ. **135**, 474).

- The Sun apparently exhibits two surface oscillations, which are quoted (Frank Hill, Allen's Astrophysical Quantities p. 343, 346, Cox Editor, 4th Edition, 1999) to have periods of 5 and 160 minutes with velocities of 10 and 50 cm/s respectively.

We shall assume that the solar oscillations can be analyzed by means of simple harmonic motion (SHM) ("University Phys." 6th Edition p. 226) where the

$$\text{Period } T = 2\pi\sqrt{\frac{m}{k}} \quad \text{.........} (1)$$

$$\text{and the maximum velocity } v_{max} = \sqrt{\frac{k}{m}} \times A_{max} \quad \text{.......} (2)$$

where A_{max} is the maximum amplitude of the oscillation, k = force constant

Since the Sun displays two distinct oscillations, each one will be evaluated separately. The total mass (m) of the subsurface nuclear complex Q is 3.3×10^{23} kg.

Case 1)

$$v_{max} = 0.1 \text{ m/s and } T = 300 \text{ sec}$$

Substituting numerical values into Equations (1) and (2) and solving for *A* and *k* gives the following:

$$A_{max} = 4.77 \text{ m}$$

$$k = 1.4 \times 10^{20}$$

Then, from Hooke's Law, $F_{max} = kA = 6.6 \times 10^{20}$ N

Case 2)

$$v_{max} = 0.54 \text{ m/s}$$

$$T = 9.6 \times 10^3 \text{ sec (160 minutes)}$$

Substituting numerical values into Equations (1) and (2) gives

$$A = 825 \text{ m}$$

$$k = 1.4 \times 10^{17}$$

which from Hooke's Law, $F_{max} = kA = 1.16 \times 10^{20}$ N

In the first case, the oscillation has an excursion of about 5 m, and the second case has 825 m. For both cases, the maximum forces exerted during the oscillation are comparable within a factor of 5 (despite their different amplitudes.)

The calculated maximum forces represent major excursions during their oscillation.

Our problem is to find the origin of these forces. We know that in order for the nuclear complex **Q** to maintain its subsurface location, its gravitational attraction of 9.0×10^{25} N has to be balanced by an equal and opposite Coulomb repulsion force in order to provide the necessary equilibrium. This force by itself is too large to account for the oscillation. There is, however, an additional smaller force produced by a Recoil Poynting Vector Force (RPVF).

The potential for a recoil force resulting from this RPVF had never been considered since no solid surface exists in the old, erroneous gas-based Sun with which it could have interacted. The matter **Q**, on the other hand, is made up of solid matter, against which the RPVF produces an equal but opposite recoil force just below the Sun's surface. **Q**, being suspended between two opposing forces, is therefore affected by this recoil force in the form of observable surface oscillations. However, the strength of the RPVF force is reduced as a function of depth penetration. As a result, **Q** gets pushed back upwards by the electric repulsion force from the core, and the downward push repeats as the maximum strength of the RPVF is reestablished. Small components of both gravitational and Coulomb forces supplement the Poynting Vector force. Since Solar gas density and its energy generation efficiency are both depth-dependent, **Q**'s relative location at and below the Sun's surface affects its fission energy output as well. Consequently, the Sun's output is controlled by this feedback phenomenon, and its constancy is assured. A Recoil Poynting Vector Force (RPVF) is (S/c) in units of Newtons per sq. meter where S is the average energy radiated and c is the speed of light. If the Sun's surface were a solid radiator, its RPVF would be $\frac{4 \times 10^{26}}{3 \times 10^{8}} = 1.3 \times 10^{18}$ N over its entire surface. In summary, the complex oscillating feed-back system thus controls the solar Luminosity and maintains its constant output.

[Based on Muller's (Allen's Astrophysical Quantities, p. 364) estimate of 5 million granules and an assumption of an equal number of corresponding subsurface SNFRs, their total surface area is less than but comparable to the Sun's surface area].

IV. The Sun's SNFR Complex

The total SNFR complex maintains its location close to the Sun's outer surface by an electric repulsion force between the Sun's negatively charged core and SNFRs equal electric negative charge, of 7×10^{16} Coulomb, counterbalanced by an equal magnitude gravitational attraction of 9.05×10^{25} N. This unique configuration ensures a time independent equilibrium of all SNFRs. Since the numerical products of their core and **Q** masses, as well as the product of their charges remain constant, (irrespective of their used and unused power consumption), the Sun's energy production rate remains constant as a function of elapsed time. Its Luminosity has therefore remained approximately constant for 4.5 billion years. *Sunspots are critical nuclear fission reactor components whose regular replacement are part of the system that keeps the Sun's Luminosity constant.*

[In a prior publication [9], (This book's first edition) the abbreviation SNFR (spinning nuclear fission reactor) was defined as NFT (active nuclear component) rather than SNFR.]

Does evidence exist for the presence of Uranium in stars other than our Sun? The answer is absolutely yes.

Two typical examples:

- "Uranium Lines in the Spectra of Peculiar a Stars," Crowley, Charles & Adelman, Saul ApJ Letters 197, 16, 5

Also note a reference in this paper to magnetic Ap stars in which ionized Uranium spectral lines were discovered.

- "First Measurement of Uranium/Thorium ratio in a very old Star," Cayrel, R., et al., "2001" Astrophysical Ages and Time Scales ASP Conference Series Vol, TBD, 2001 T. von Hippel, N. Manset, C. Simpson

V. The Corona

Based on our knowledge of commercial reactors, the Sun's SNFRs are likewise producing atomic elements at elevated temperatures as well as slow and fast solar wind, consisting of neutrons. which morphed into protons and electrons. SNFRs also produce a range of high velocity, ionized, nuclear particles, (as reported by [3]) plus high-energy, ionized fission fragments, and gamma rays, all interacting with the Corona's plasma and thereby establishing its inferred 1–2 million effective temperature [3].

Figure 1 *is a vivid demonstration of the active nuclear fragments on the Sun's surface of a photo taken in ultraviolet light.*

Figure 2 *is a photo of the South Pole side of Jupiter showing subsurface granular features without much visible activity. At a cooler temperature, their past activities resulted in the production of colored molecular substances.*

The Sun's solar wind (primarily its protons) will eventually interact with the Earth's magnetic field and produce the spectacular auroras. Some solar wind may simultaneously impact LIGO's unshielded two ultra-sensitive detectors and be mistakenly identified as gravity waves. A 600 ft underground (Ligo type design) detection system (the Japanese Karagra) would not be able detect solar wind protons, and thus make it even far more unlikely to detect non-existent gravity waves.

VI. The General Relativity Connection

There were two successful applications of General Relativity, both of which involved the Sun.

A) The enigmatic advance of the planet Mercury's perihelion. (Appendix A)
B) The bending of electromagnetic waves grazing the Sun's surface.

General Relativity's correct solution provided the crucial evidence for the Sun's subsurface hitherto fore unknown mass Q. An alternative approach to this problem assumed that Mercury experienced a minute increase in speed during its hundred-year revolution about the Sun. It also elicited the unusual phenomena of an unexpected departure from the mass-equivalence principle due to Q's gravitational attraction being balanced by an exactly equal electrostatic force.

B) Deflection of Electromagnetic Radiation Near the Sun

The bending of starlight grazing the Sun's surface, can readily be explained as being due to the sum of the Sun's gravitational field *and* the Sun's surface electromagnetic field, both being precisely equal in magnitude. A simple classical Newtonian physics calculation predicts a bending of such radiation by an amount $2GM/Rc^2$, where G, M, and R are the values for the gravitational constant (G), mass of the Sun (M) and its Radius (R), c being the speed of light. Also, that photons have a mass given by hv/c^2. In the field of physical optics, light is bent within dielectric media upon entering and leaving a glass prism. It is the electric field component within a dielectric medium that interacts with the photons. In a similar fashion, it is proposed here that distant starlight, or any electromagnetic waves, grazing the Sun's surface are bent by the Sun's gravitational field *plus* the electric field component at the Sun's surface. If electric and gravitational forces are equal at the Sun's surface, the electric force contribution of the electromagnetic wave bending must also equal $2GM/Rc^2$. Thus, the total angular bending is the sum $4GM/Rc^2$. Experimental verification of this theoretical value is well-documented in the literature (beginning with Eddington's 1919 solar eclipse expedition). This verification can also be construed as confirmation that the Sun's surface electric field and the gravitational components are equal.

Corollary

- The inferred mass of Q is exactly equal to the mass of Mercury. This unexpected result requires special consideration. It implies:
  a. Most likely, the Sun has been producing and maintaining identical amounts of Q at this level, for billions of years.
  b. See other possible implications related to a novel solar system development theory (*Voyage into Astronomy* 2015).

References

[1] E.N. Parker, *A Critical Review of Sun-Space Physics,* Astrophysics and Space Science **277,** 1 (2001)

[2] E.N. Parker, *Exploring the Innermost Solar Atmosphere,* Nature Astronomy **4**, 19 (2020)

[3] S.R. Habbal et al. *Identifying the Coronal Source Regions of Solar Wind Streams from Total Solar Eclipse Observations and in situ Measurements Extending over a Solar Cycle,* ApJ.Letters **911**, L4 (2021)

[4] F.M. Johnson, *Planetary Cores, their Energy Flux Relationship and its Implications,* Planetary and Space Science, **151,** 125 (2018).

[5] L. Motz and Duveen, *Essentials of Astronomy,* Watsworth Publishing C., Inc, (1966), p445

[6] H.A. Bethe, *Energy production in Stars* Phys.Rev. **55,** 434 **(1939).**

[7] F.M.Johnson, *Voyage into Astronomy,* Second Edition, Lawyers and Judges Publ. Co, Inc.(2015).

[8] R.Muller, *Allen's Astrophysical Quantities,* A.N.Cox Editor (1999), p.364

[9] F.M.Johnson *recent Discoveries in Astronomy (Forensic Astronomy),* Lawyers and Judges Publishing Co.(2021)

Appendix

Mercury's Orbital Perihelion Advance

Mercury's orbital perihelion advance involves its elliptic orbit as a whole rotating slowly about the Sun. This phenomenon is only obvious because Mercury's orbit is an ellipse rather than circular. It results from perturbations by the rest of the planets acting on Mercury. Classical mechanics is able to explain roughly 5600 seconds of arc per century. However, it leaves 43 seconds of arc unaccounted for. This is where General Relativity was applied successfully to explain it.

The detailed calculation below is not a substitute for Einstein's correct General Relativity analysis. It is merely a demonstration that an assumed, hidden sub-surface **mass (Q = 3.3 × 10²³ kg)** would have been detected by Newtonian physics in the absence of an opposing electric repulsive force on a mass Q (This would have increased Mercury's orbital speed by 4.4 mm/s out of 47 km/s). For mass Q to remain in its present location in the assumed absence of a repulsive electrostatic force, it would need a substantial speed increase to remain in the Sun's outer surface orbit. Its presence at the Sun's subsurface would then have generated an additive force of $T = 2.169 \times 10^{15}$ N, between planet Mercury and itself, a force which would have been readily detectable by Newtonian physics. The calculation below attempts to calculate this force T. It is not a substitute for General Relativity!)

Necessary Parameters:

$P = 7.607 \times 10^{6}$ seconds (Period of one revolution of Mercury)

$a = 5.790 \times 10^{10}$ m (semi-major axis of Mercury's elliptic orbit)

$G = 6.672 \times 10^{-11}$ in mks units

$M_\theta = 1.989 \times 10^{30}$ kg (mass of Sun)

$v = 4.787 \times 10^{4}$ m/s (velocity of Mercury)

$R_\theta = 6.955 \times 10^{8}$ m (radius of Sun)

$L_\theta = 3.845 \times 10^{26}$ Watts (luminosity of Sun)

$m_{Hg} = 3.302 \times 10^{23}$ kg (mass of Mercury)

$c = 2.9979 \times 10^{8}$ m/s (speed of light)

$\Delta\theta$ = the advance of Mercury's perihelion (radians/yr) (360° = 2π radians)

Mercury performs 415.28 revolutions in 100 years.

There are 206265 arc-seconds in 1 radian.

Using Newtonian Physics, the speed of Mercury (Hg) is derived as follows:

$$\frac{mMG}{a^2} = \frac{mv^2}{a}$$ for a circular orbit [using the Principle of Equivalence] .... (1)

$$\text{Then } v = \left(\frac{MG}{a}\right)^{\frac{1}{2}} = 4.787 \times 10^{4}\,\text{m/s} \qquad (2)$$

Now assume an **extra** hypothetical force **T** is operating.

$$\text{Then } \frac{mMG}{a^2} + T = \frac{mv^2}{a} \qquad (3)$$

$$\text{Solving for } v = \left[\frac{MG}{a} + \frac{Ta}{m}\right]^{\frac{1}{2}} \qquad (4)$$

The extra term represents the increased velocity (v_i) due to force T. Thus:

$$v_i = \frac{Ta}{2vm} \qquad (5)$$

Extra distance travelled due to T is:

$$\Delta s = v_i P = \frac{T \cdot Pa}{2mv} \qquad (6)$$

The angular measure of the so-called "perihelion advance" (during the course of one revolution of Mercury) is:

$$\Delta\theta = \frac{T \cdot P}{2mv}\ (\text{radians}) \qquad (7)$$

Calculation of T

The Sun would exert a small ***additional*** gravitational force (T) due to Q's "pseudo extra" mass m_b (3.3×10^{23} kg), which is situated close to the outer solar surface. This is alluded to above.

$$T = \frac{Gm_b m_{Hg}}{r^2} \quad \text{.......} (8)$$

Where:
$\mathbf{m_b = 3.3 \times 10^{23}}$ **kg**
m_{Hg} = mass of Mercury

$$T = \frac{(6.672 \times 10^{-11})(3.3 \times 10^{23})(3.302 \times 10^{23})}{((5.790)^2)(10^{20})} \quad \text{......} (9)$$

$T = 2.169 \times 10^{15}N$ .... (10)

$$\Delta\theta = \frac{T \cdot P}{2mv} = \frac{(2.169 \times 10^{15})(7.607 \times 10^6)}{(2)(3.3 \times 10^{23})(4.787 \times 10^4)} = 5.22 \times 10^{-7}$$

.....(11)

$$\Delta\theta = (5.22)(10)^{(-7)}(415.28)(2.06265)(10)^5$$

seconds of arc in 100 years .... (12)

$\Delta\theta$ in 100 years → (0.99)(44.7) = **44"** (±1') .... (13)

(0.99 factor is needed because of Mercury's elliptical orbit)

[Latest astronomically deduced value for $\Delta\theta$ = 43"]

Planet Mercury (m_{Hg}) is utilized here as a long-distance probe of the Sun.

By assuming the correct value for the advance of Mercury's perihelion to be 44 ± 1", the predicted mass of ***Q*** would then be $\mathbf{3.3 \times 10^{23}}$ **kg.**

Chapter 3

Solar System Evolution and Its Implications

Abstract
A recent discovery of the Sun' s fission energy production mode, required a detailed reexamination of both the sun and its planets evolution scenario. As a result, the presence of a Planetary Nebulae was discovered. The double ring Solar Nebula consists of multi- million spinning nuclear fission reactors (SNFRs) with speeds of 2.33 km/sec. Their large rotational kinetic energy enables each one to launch hundreds of spinning comets, which in turn produce atoms and molecules, revolving at speeds in the 350 -400 km/sec range, about each comet. Among these spinning molecules is the most thermodynamically stable, fluorescent molecule, MgTBP, whose lowest energy level emits fluctuating, blackbody radiation at 2.725K. Due to these molecule's revolution about their comets, a 368 km/sec Doppler velocity correction is required, prior to being able to study the Cosmic Microwave Background. Similar fluctuating red fluorescence at 6284 Å or 6350 Å is predicted.

1. Introduction
It is a well-recognized challenge to undertake an extended scientific analysis concerning the evolution of our 4.5 billion years old solar system. Yet, there are an abundant number of relevant clues: a) The Bode-Titius law of 1772, b) precise planetary distances relative the Sun. [1] c) orbital velocities of each planet [1] d) their approximate commonality of orbiting in the plane of the ecliptic and in a similar sense of circulation, with the exceptions of Halley's comet and some of the planetary satellites, [2] e) the recently discovered nuclear fission within the Sun and all of its planets, [3] f) planet's eccentricities. [4]

In addition, there is a collection of data that at first glance appears to be unrelated. These include the Sun's energy production mode [5]. Its dense spinning nuclear core, its multimillion spinning nuclear fission reactors (SNFRs) below the Sun's surface, their total mass being equal to the mass of planet Mercury [6], The important role of Sunspots. Finally, based on a conservation of angular momentum analysis [7], our Sun evolved from a dense Neutron sphere of mass 97 times our present Sun's mass; a value confirmed in Fig.1 of ref. [5].

2. Evolution of Solar System
The solar system development began by successive launchings of millions of SNFRs from a rapidly spinning (97 M_0) Neutron core (B), into a coplanar 0.4 A.U. orbit [7]. The process involved a complex hybrid combination of electrically charged spinning nuclear fission reactors (SNFRs), gravitational, electromagnetic, and nuclear forces. The core flips 180 degrees and launches a second set of multimillion SNFRs, emanating from its semi-hemispherical outer nuclear shell surface into its 0.4 A.U. orbit. Based on the core's mass, ejected SNFRs have a calculated speed of 4.6×10^5 m/sec. These first two launchings are unique and require special consideration: Subsequent to the launchings of eight planets, they become our "Solar Planetary Nebula": A set of rings containing millions of electrically charged SNFRs, with speeds in the 2.33 km/s range, at 80 A.U. distances from the nuclear core.

The massive 97 M_0 core undergoes its own development by breaking off a smaller size spinning core of mass

M_0 (which later becomes our Sun), designated **A,** while the larger, spinning progenitor core is referred to as **B.** Their role in subsequent planet development phases is inferred via each planet's known eccentricity [4]. Both cores revolve about their common center of gravity during launchings from core B. Table 1 lists each planet, its eccentricity and B's mass at each launch. All such launchings resulted in placing millions of SNFRs into an orbit 0.4 A.U. distant from the cores. Planet Neptune was the first one in the launching sequence, while total core mass was (48 + 1) $M_0 = 98 \times 10^{32}$ kg. After each planet's launch, core B splits in half, with one of its spinning halves disintegrating into multiple smaller spinning components that are ejected symmetrically into the Interstellar Medium ISM [7]. Fifty percent core mass reduction results in a comparable decrease of gravitational attraction and thereby the doubling of Neptune's distance.

With the exception of rings 1and 2, all of the other millions of spinning NSFRs, launched into each individual orbit, amalgamate and merge to produce a single spinning planet. Thus, each planet in our solar system was the result of a similar development sequence. Asteroids are notable exceptions. They should have become a planet. However, SNFRs were gravitationally perturbed due to their proximity to the planet Jupiter, which distorted their orbits and thus prevented their merger. Some individual Asteroids are still rotating and display comparable SNFR sizes.

Another development example is the planet Mercury. Its mass is exactly equal the total mass of the Sun's subsurface multimillion SNFRs [6], which implicates the proto-Sun in producing its origin.

Planet Neptune development is sequentially duplicated by each planet. After each successful planet launch, distances of all planets are doubled due to a 50% core mass disappearance subsequent to each launch. This raises the question as to how the 0.4 AU launch distance can be maintained if its gravitational attraction is comparably diminished. However, all SNFRs are electrically charged as well as the nuclear cores. Their electrostatic repulsion is balanced by their gravitational attraction, since both forces are governed by an inverse square law distance relationship, the 0.4 AU distance remains constant, provided their mass to charge ratio remains constant.

Table1

Solar system objects, listed in order of their evolution, starting with the Planetary Nebula of Rings 1 and 2.

	Core mass at launch		Distance from core	
	(2×10^{30} kg)	eccentricity	Bode-Titius (A.U.)	Actual
Ring 1	96+1	0	76.8+.4=77.2*	80
Ring 2	96+1	0	76.8+.4=77.2*	80
Neptune **	48	.008	38.4+.4=38.8	**39.4
Uranus	24	.047	19.2+.4=19.6	19.1
Saturn	12	.059	9.6+.4=10	9.5
Jupiter	6	.048	4.8+.4=5.2	5.2
Asteroids	3	—	2.4 +.4=2.8	—
Mars	1.5	.093	1.2+.4=1.6	1.5
Earth	0.75	.017	0.6+.4=1.0	1.0
Venus	(1.5)***	.006	0.3+.4=0.7	0.72
Mercury	1	.205	0+.4=0.4	0.39

* Extrapolated values
** Neptune's original distance prior to a rearward mass expulsion
*** extrapolated core mass based on eccentricity-see text for its derivation.

3. Origin of SNFRs

How is Uranium 235,238 produced in a spinning nuclear fission reactor?

SNFRs originate from a rapidly spinning **neutron** core, which favors the production of atoms having the largest number of neutrons. This in turn produces a large number of diverse, stable and unstable atoms, including Uranium 238 and 235, plus fast and slow neutrons. Thus, fission reactors originate directly on the Sun's spinning core surface. Due to the solar core's exceedingly high spin rate, neutrons at the core's equator become preferentially ejected. This results in the separation of the core's surface into two distinct surface area halves. Subsequent to their ejection from the core, they constitute hollow SNFRs, when they eventually arrive at the Sun's surface. (These are the *sunspots*)

Beta decay produces protons and electrons. Alpha decay, from large Z and A atoms, produce helium. Thus, the preponderance of hydrogen and somewhat lesser quantities of helium atoms, observed in stellar surfaces, are accounted for.

4. Planets Venus and Mercury

The end of the solar system evolution is unique: The planets Venus and Mercury were launched at approximate equal distances from their core origin: Venus was launched by core B and Mercury by core A.

Venus ended up in an approximate circular orbit, (e=0.0068), its orbit inclined 3.4 degrees with respect to the ecliptic. Mercury's eccentricity is 0.2056 degrees, its orbit inclined at 7.0 degrees, (consistent with the Sun's inclination with respect to the ecliptic). The total mass of the double central core was (2.0+1.86) M_0.

Mercury was launched exclusively from core A while revolving about a common center of gravity of both cores. This is deduced from Mercury's highly eccentric orbit (e=0.2056). Core B carried an electric charge which ensured the constant, approximate 0.4 AU distance, since Venus was controlled by both gravity attraction and electric force repulsion. In fact, all planet launchings involved a force balance of electrically charged SNFRs and electric core repulsion being equal to their gravitational attraction.

The final disappearance of remnant **B,** resulted in Venus' orbit to move out to 0.723 AU, due to both a diminished central core mass and electric charge. This however did not affect the planet Mercury.

5. Detailed launch sequence

Merging was expedited via alternating SNFR spin directions, which produce mutual magnetic attraction of adjacent SNFRs. This feat was accomplished by core **B** spin flips, subsequent to every hemispherical nuclear surface shell ejection from its fast spinning core. The sole exception to this procedure were the first two launchings which involved non-coalescing electrically charged SNFRs.

Core **B'**s angular momentum was sufficiently large to ensure that launched planets could achieve dynamic equilibrium in their respective orbits, based on the gravitational attraction to the combined mass of cores **A** **&B**, Whereas planet Mercury was the only planet launched exclusively by core **A**.

Except for planet Mercury, all planet launchings and associated core splitting originated from the spinning core **B**. Gravitational control functions became core **A**'s responsibility after **B'**s departure.

6. The Sun's Development

The Sun resumed its electrically charged SNFR production by launching them in an orbit consistent with its available core angular momentum and electric charge, into its present radius of 7×10^8 m. Its core spin directions reversed after each launch which resulted in alternating SNFR spin directions. Thus, SNFR spacings were balanced by electric charge repulsion and magnetic attraction.

By having about 5 million SNFRs spread over the entire Sun's subsurface and heating its gas content, radiant energy of about 4×10^{26} watts began to be generated. Burned out SNFRs are periodically replaced by fresh ones, first detected by Galileo's telescope in 1610 and since then studied as "Sunspots". Profuse neutron evaporation, from their spinning "virgin" surface, results in their lower surface temperature, compared to the average, overall, Sun's surface temperature.

7. SNFR Properties

All SNFRs are hollow with internal gas under pressure. During their merge accumulation phase to form each planet, the bulk of their total accumulated gas content is substantially retained by each (and within) the four giant planets, except for a glitch in planet Neptune. Its internal gas pressure was excessive causing a gas and mass ejection in an approximate rearward direction, thereby increasing Neptune's velocity [7}. This caused it to move from a solar distance of 39 A.U. to 30 A.U. Its old position is replaced by the ejecta of Pluto/Charon and other minor planetary objects. Planets Mercury, Venus, Earth, and Mars lost their lightweight gasses due to insufficient gravity. (Table 1 provides a summary for key planet properties.)

Subsequent planetary satellite developments relieved excessive planetary core angular momenta in all planets with satellites.

8. Implications

To repeat: Our solar system development scenario began with the launchings of two sets of electrically charged SNFRs, originating from a rapidly spinning, electrically charged 97Mo mass core, into a 0.4 A.U. distant orbit. Its multimillion, *non-merging*, SNFRs had an initial speed **466** km/sec. Subsequent to the launching of the first set, the nuclear core flipped 180 degrees. Thus, all SNFRs of the second set are spinning in the reverse direction while moving in the same anticlockwise direction as SNFRs of the first set. At the conclusion of all planetary launchings and its associated, substantial core mass reduction, the double ring SNFRs were left at 80 A.U. distant from the Sun's core. According to the principle of "conservation of angular momentum", all SNFR speeds are reduced by a factor of (80/0.4) to **2.33** km/sec., at the conclusion of planet launchings.

The average speed of the second set of SNFRs (in ring #2) is less than that of the first set, due to a slight core mass reduction, subsequent to the launch of ring #1.

At 80 A.U, their individual, larger separation, plus rotational kinetic energy retention, enables each SNFR to launch hundreds of satellites/comets into orbit.

Comets in turn produce and eject a multitude of diverse organic molecules and atoms, from their internal nuclear power source, via a plasma jet stream [8]. While the bulk of the ejecta remains close to the ecliptic plane, a small quantity does venture out of the ecliptic, eventually spreading out over 4π steradians. The significance of the entire planet-evolutionary scenario is the following discovery: **Our entire solar system is surrounded by a complex Planetary Nebula.**

9. Planetary Nebula in our galaxy

This presents an opportunity for comparison with approximately 2000 known planetary nebulae in our galaxy. The main difference being that we are inside and therefore unable to observe any absorption spectra from a multitude of atoms and molecules within our nebula. However, our Sun is close enough to our nebula to excite its atoms and molecules, which are capable to fluoresce.

At solar system's evolutionary conclusion, our Planetary Nebula had a diameter of about 0.0024 light years, which in 4.5 billion years would increase in size, comparable to other nebulae. Each Planetary Nebula in our galaxy also displays a very hot central star with coronal temperatures in the million-degree range. Their rarity and relative short lifetimes are consistent with the general concept of other solar system beginnings. The following are direct quotations from a Palomar Observatory/Cal Tech publication, whose observations are comparable to our own solar systems evolution: Viz.:

"There are usually some ***brightness fluctuations*** *over the disc. The planetaries generally have regular inner boundary as well, giving them the appearance of a* ***ring***. ***Many have two lobes of bright material****, resembling arcs of a circle, connected by a bridge, somewhat resembling the letter Z."*

"The (nebular) expansion is proportional to the distance from the central star, consistent with the entire mass of gas having been ejected at one brief period from the star in some sort of instability".

"Planetary nebulae are chemically enriched in elements produced by ***nuclear*** *processing within the central star."*

Our solar system's Planetary Nebula also has two spinning "**rings**" of multi-million spinning SNFRs in each ring. The speeds of the first set of SNFRs is larger by factor of (1 + 1/360) relative to SNFRs in the second ring. Each of the spinning SNFRs undergoes its own development by launching thousands of satellites/comets. Each spinning comet produces a host of atoms and molecules which are ejected via a plasma jet stream [8]. The Sun's impinging, blue color photon radiation, towards this vast collection of molecules, excites the most thermodynamically stable molecule: MgTBP=(Magnesium-Tetrabenzoporphyrin) [9]. Its red fluorescence at 6284 Å or at 6350 Å should be observable in the same spatial locations as its lowest vibrational wavelength, corresponding to the CMB temperature of 2.725 K.

10 NASA's COBE and WMAP satellites

The vast amount of excellent, accumulated data from NASA's COBE [10] and WMAP [11] plus Planck' surveyor ESA space missions, is herewith reanalyzed as a kinematics physics problem of our early solar system's evolution.

Specific, spinning molecules, associated with the two rings, emit Black-Body Microwave Radiation in the ecliptic plane. This BBMR also overlaps part of the Milky Way galactic region. It is essentially ignored in favor of the greatly diluted outlying signals, in order to elicit tiny fluctuations in the MBR. A Fourier transform decomposed a sky map into spherical harmonics resulted in a power spectrum revealing a dipole anisotropy in the first spherical harmonic. The largest power fluctuation occurred at scales of about one degree at a MBR amplitude of about 0.00336 K. Further multipole moment analysis revealed four additional smaller size peaks (see Fig.1) whose total power is exactly equal to that of the first primary peak (a), at the one-degree spatial resolution mark. The key to understanding these peaks is the recognition that the smaller, equal size peaks from ring 1 [(b) and ring2 (c)], arise from microwave energy reflections, towards the earth, of spinning MgTBP molecules. As they revolve about the solar system's center of gravity, the

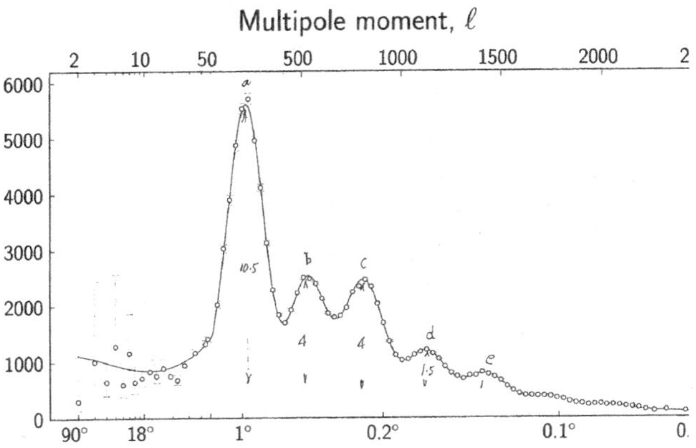

***Figure 1** Data derived from the WAMP science team of a power spectrum revealed dipole anisotropy in the spherical harmonics of a Fourier transform decomposed sky map. Multipole moment analysis of the 2.723K Cosmic power fluctuations could separate signals from each ring. Their sum is precisely equal to that of the largest, at the one-degree mark, as a consequence of a fortuitous overlap of Ring speeds 1 & 2.*

differential (1 + 1/360) ring speed, between rings 1 &2, causes a spatial overlap of fluorescent molecules from each ring, resulting in their energy summation at the one-degree revolution mark. Each of the 4 peaks represents a different set of molecular speeds of revolution. The equal sized (b) and (c) peaks exhibit speeds of individual fluorescent molecules from each ring, revolving about their comet's center of gravity, while emitting 2.725K blackbody radiation. (see Fig.1)

11. Details of the two-ring launch sequence

The original core mass of $97M_0$ launched its multimillion SNFRs twice, at a distance of 0.4 A.U. from its core, at a speed of 465.2 km/sec [based on equation $V=(MG/R)^{1/2}$]. Subsequent to the launching of eight planets, the two rings moved to a distance of 80 A.U. Consequently, the original SNFR speeds were reduced by a factor of 80/0.4 resulting in a speed of 2.326 km/sec. The time for one complete ring revolution $(P)=2\pi R/V=3.23073\times 10^{10}$ sec. The speed of ring #1 however, is larger than ring #2 by a factor of (1.0027778), viz. (1 + 1/360). Thus, $V_1=2.33246$ km/sec and $V_2=2.32600$ km/sec.

The gravitational mass (M_1) which controls ring1 is: $M_1=R\,V_1^2/G=0.97517\times 10^{30}$ kg

The gravitational mass (M_2) which controls ring2 is $M_2=R\,V_2^2/G=0.96983\times 10^{30}$ kg

The two-ring system sum (M_1+ M_2) requires a central core mass of 1.945×10^{30} kg to keep it gravitationally in place. Since the Sun's mass is 1.989×10^{30} kg it is clearly our Sun which controls the double ring system. **It also confirms all prior assumption regarding ring distance and their extrapolated velocities.**

The mass of ring 1 is (M_1 minus M_2)$=5.34\times 10^{27}$ kg, consistent with the mass of the heaviest of our planets.

12. Temperature and special differences of the two-ring system

Differential speeds of the two-ring structure can be related to a reported WMAP spectral temperature difference of $\Delta T=0.0002$ K., in the CBR as follows:

Interpret the so called CBM fluctuations as arising from two locations, separated by a relatively small distance ΔL and a corresponding temperature difference ΔT.

The blackbody Intensity $P=Area\times\sigma T^4$, and $\Delta P=4\times Area\times\sigma T^3\,\Delta T$, hence

$\Delta P/P=4\times\Delta T/T$ Also, since P obeys the inverse square law, $\Delta P/P=-2\,\Delta L/L$

Substituting numerical values: T=2.725K, L=80 A.U., and $\Delta T=0.0002$ K.

$\Delta P/P=0.0002936=2\Delta L/L$Thus,$\mathbf{\Delta L=1.17\times 10^{-2}}$ **A.U.= $\mathbf{1.757\times 10^6}$ km**

13. Origin of the fluorescent MgTBP molecules [9]

The two revolving rings with its moving multimillion SNFRs are 1.757×10^6 km. apart. Each SNFR launches a set of spinning comets, and each comet produces a large multitude of organic molecules, from its hot interior, by a plasma jet process [8]. Among this large array of molecules will be the most stable, fluorescent molecule MgTBP [9], which can tolerate the highest temperature and in fact sublimes, making it virtually indestructible. Moreover, MgTBP is either alone or surrounded by a set of octanes. Solar radiation impinging on MgTBP causes it to fluoresce at either 6284 Å or at 6350 Å dependent whether alone or surrounded by hydrocarbons. In either case, its lowest energy level is distributed in the molecule or its enveloping structure to produce the blackbody 2.725K spectrum the so-called CBR (Cosmic Background Radiation). Thus, CBR signals arise exclusively from MgTBP's fluorescence and their association with a combination of two separate ring rotations, comet revolution about SNFRs and atom and molecule revolutions about comets. Their combined microwave fluorescent (Fig1) radiation appears via a detailed angular resolution study which produced a power spectrum map showing the largest microwave fluctuations at the one-degree mark. The one-degree mark in turn is a result of the two-ring differential velocity.

While the bulk of the microwave 2.725K microwave background radiation is in the ecliptic plane, a small percentage of SNFRs had migrated out of the ecliptic and eventually became scattered over 4π steradians. First NASA's COBE and then its WMAP satellite enabled a more detailed, finer angular resolution study of this wider region, thereby discovering fluctuations in the (CBR) microwave background. Furthermore, a so called "dipole anisotropy" (misinterpreted as solar motion) was discovered to be superimposed on the CMB. which required removal prior to a fluctuation analysis.

14. NSFR spin flips

In both rings, multimillion NSFRs revolve counterclockwise, at speeds in the 2.33 km/sec range. However, a 180 flip-over of NSFRs in the second ring resulted in their spinning in a clockwise direction, while still revolving in an anticlockwise orbit about the Sun. Atoms and molecules ejected from the spinning comets follow the spin directions of each comet. (Among the multitude of ejected molecules will be the unique, red fluorescent molecules MgTBP). SNFR's rotational kinetic energy is successively transferred to its spinning comets and then to all ejected molecules, with velocities of revolution about each comet in the 350-400 km/sec range. These velocities are large enough to require either a blue or red shift Doppler correction for each of the spinning MgTBP molecules.

15. COBE and WMAP science team's sky map

The COBE [10] and WMAP science team [11] published a sky map, using three false color images of the sky, as seen at microwave frequencies. Our interpretation of their "yin-yang" pattern, using a Doppler wavelength analysis, is consistent with a velocity (v) of MgTBP molecules revolving about their comet origins at approximate 368 km/sec.

Doppler equations can be simplified using Wien's Law ($\lambda T = 0.29$ cm K) to read:

$T_0/T = (1-v/c)$ or $T_0/T = (1+v/c)$ for a blue or red shift respectively, where $T_0 = 2.725$ K and the false colors from the WMAP report are given as either $T = 2.729$ K or $T = 2.721$ K respectively.

The results from NASA's WMAP spacecraft analysis [11] are shown in Fig1. In summary, the angular power spectrum is fully explicable, for each one of the five peaks, based on the above analysis. The so-called fluctuations are all, in fact, **microwave emissions from revolving MgTBP molecules**. Note that emissions from each ring are clearly distinguishable: peaks (**b**) and (**c**) arise from ring **1** and ring **2** respectively while peaks (**d**) and (**e**) do likewise. A lineup of all four peaks produces the maximum signal at the one-degree mark (**a**) at a signal magnitude equal to the sum of the other four. The largest of the frequencies (b and c) required the above Doppler frequency corrections, prior to extracting the 2.725K blackbody spectrum.

16. Extreme Trans-Neptunian Objects (ETNOs) [12]

About half a dozen objects have been identified as ENTOs orbiting our Sun. Among them was "Sednoid 2012 VP113" whose perihelion distance is **80 A.U**.

Other such ENTOs are "Nika"2011 KT11 and "208 KV42" whose orbital inclinations are 110 and 104 degrees respectively, with respect to the ecliptic. The latter object is in a retrograde orbit about the Sun. Since the two-ring separation is relatively small, it is conceivable that interaction between some satellites of adjacent orbits could interact to perturb their motion. The retrograde ETN object had to have originated from ring #2, since its SNFR's spin is in the reverse direction from those in ring #1.

17. Conclusions

A revolving double-ring/belt, a Planetary Nebula at 80 A.U., encloses our Solar System. It consists of millions of SNFRs in each ring, each with a multitude of revolving comets (satellites) and a huge revolving conglomerate of atoms and molecules. Among this multitude is an ultra-stable, spinning molecule (MgTBP) that absorbs solar input, and as a result of its high velocity spin about its comets, fluoresces fluctuating microwave radiation at 2.725 K.

It is herewith predicted that this unique molecule (MgTBP) should also emit **identical, <u>fluctuating</u> radiation at either 6284Å or 6350 Å.**

Similar solar system and Planetary Nebula developments are presumably common throughout the universe. Consequently, the ubiquitous occurrence of atoms and molecules in the ISM point to a common origin. Moreover, a relationship most likely exists between the presence of over 500 Diffuse Interstellar Bands (DIBs) in the ISM and their unambiguous MgTBP identification as the DIB carrier [9] and [6] Chapt.8.

A NASA submillimeter microwave measurement of MgTBP powder [9] produced an absorption line at 341 Ghz, from which its equivalent blackbody temperature of 2.725 K can be deduced. (See appendix).

The Real Origin of the CBR

Extensive spectroscopic laboratory studies have identified Magnesium tetrabenzoporphyrin (MgTBP) (see Figure 2) as the primary constituent of interstellar grains.[9]. These

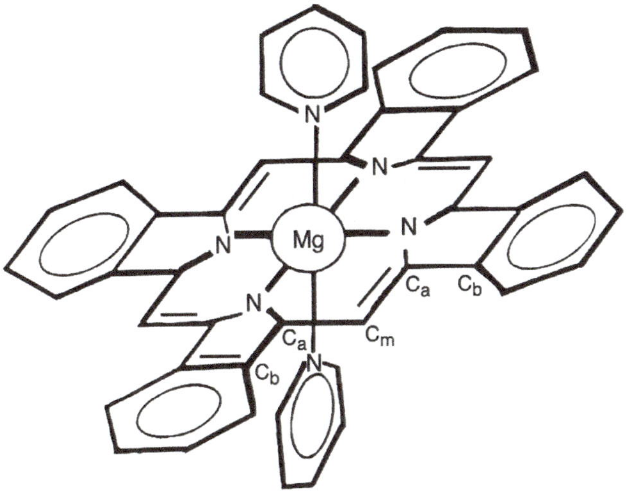

Figure 2 *Structure of MgTBP.*

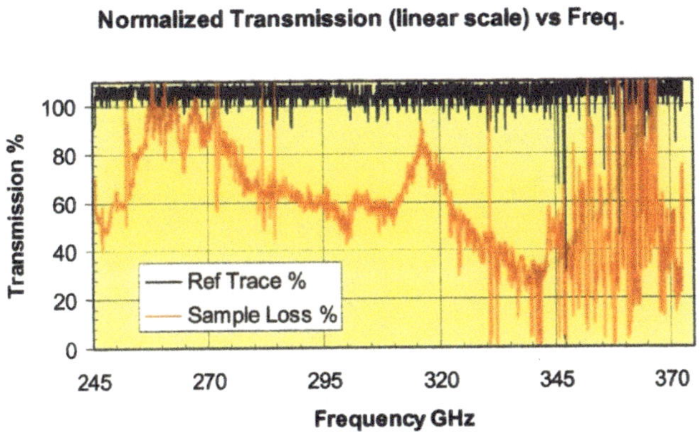

Figure 3 *MgTBP powder microwave absorption spectrum, showing minimum at 341 Ghz.*

grains are ubiquitous in interstellar clouds throughout our galaxy. Background stellar radiation illuminates these grains and excites their entrapped MgTBP molecules. These molecules then display their visible signatures (DIBs), release IR radiation, seen as the UIR, by heating their matrix environment, the matrix is primarily composed of an assortment of paraffins (for instance, octane, nonane, and decane), which in a frozen state, resemble fluffy snow. (Experimental details can be found in in Chapt.8)

In our solar system's Planetary Nebula, as well as in the interstellar medium, MgTBP molecules, stimulated by blue radiation, are highly efficient infra-red fluorescence emitters. Its lowest vibrational energy level, which is 11.3 cm^{-1} above ground level, has been precisely measured by NASA in a powder,[9], by means of a microwave absorption method at $\nu = 341 \pm 2$ Ghz. MgTBP molecules are generally situated within a paraffin (Shpolskii) fluffy matrix, which allows them to vibrate but not to rotate. The measured lowest vibrational frequency corresponds to the allowed bending modes of the 4 outer benzene rings and the 2 pyridine molecules attached to the molecule's central magnesium atom. There are $n = 6$ possible vibrators providing N degrees of freedom, where $N = 3n - 6 = 12$.

The lowest vibrational energy of MgTBP, measured at ($\nu = 341$ GHz) (See Figure 3), is transferred to the phonon vibrational modes of its host crystal lattice (paraffin matrix).

The laws of thermodynamics provide the following equation:

$N(\frac{1}{2}kT) = h\nu$ Where each degree of freedom N is associated with ($\frac{1}{2}kT$)

k = Boltzmann constant

h = Planck's constant

Substituting numerical values gives

$T = 2.727 \pm 0.008$ K

The latest astronomical value is 2.725 ± 0.002 K (Mather et al. 1999)

This implies that MgTBP molecules within grains would radiate as black bodies at an equivalent temperature of 2.727 K. This fact, their rotational (Doppler Effect) measurable mobility plus their extended dispersion within our solar system's, double ring Planetary Nebula, provide a simple, testable explanation for the cosmic microwave background radiation.

Addendum:

Confirmation of Solar System's Planetary Nebula by Pioneer 10

Pioneer 10, on 4/27/2002 produced the last successful reception of clean data, lasting 33 minutes of telemetry, at a distance of **80 A.U**.

Nine months later, on Jan.23,2003, final, very weak signals were received from Pioneer 10. Subsequent signals were barely strong enough to detect.

Based on the spacecraft's speed of 12.5 km/sec, Pioneer 10 had encountered atoms and molecules when it penetrated the Planetary Nebula for a distance of about 3×10^8 km.

References

[1] A.N. Cox (Editor) Allen's Astrophysical Quantities, 4th Edition, 1995, p.294

[2] A.N. Cox (Editor) Allen's Astrophysical Quantities, 4th Edition, 1995, p303

[3] F.M. Johnson, Planetary and Space Science 151 (2018) 125-129

[4] A.N. Cox (Editor) Allen's Astrophysical Quantities, 4th Edition, 1995, p. 295

[5] F.M. Johnson, Planetary and Space Science 151 (2018) 125-129

[6] F.M. Johnson, Forensic Astronomy, Lawyers & Judges Publishing Co.Inc., 2021, p.21

[7] F.M. Johnson, Voyage into Astronomy, Lawyers & Judges Publishing Company, Inc., 2ndedition, 2015

[8] G.W. Hodgson, Ann.N.Y.Acad.Sci. 206(1973) 670

[9] F. M. Johnson, Spectrochemica Acta Part A65 (2006), 1154-1179

[10] N.W. Bogges, J.C. Mather, et.al. ApJ 397 2 (2007) 377 (The COBE Mission)

[11] C. L. Bennet, D. Larson, J.L. Weiland,et al. ApJ Supplement Series 208 (2013) Article #20 ((WAMP)

[12] C.& R. de la Fuente Marcos, Mon. Notices Royal Astronomical Soc.443 (2014) L59-L63

Chapter 4

Solar System Development (Details)

The material in this chapter is also developed in the author's textbook *Voyage into Astronomy 2nd Edition* (2015).

Credit: NASA, March 22, 2017

Synopsis

The Pluto-Neptune Connection and Its Interloper

The present orbits of Pluto and Neptune, will be reexamined in order to gain some insight into their remote past.

The perihelion distance of these two planets is almost identical. (The perihelion distance is the closest distance of a planet to the Sun.) The difference in their present perihelion distance is only 0.22 A.U. Thus, these two planets could have been very close together in the past. In fact, perhaps they were once together! This scenario will now be examined: First note the famous Halley's comet as the mysterious interloper, whose significance will now be speculated, since Halley's cornet has about the same inclination to the ecliptic as Pluto, except it moves in the opposite direction: it retrogrades. Furthermore, at Halley's comet's extreme distance (aphelion) away from the Sun, it is about in the middle of Pluto and Neptune's mean distance from the Sun. What does this all mean?

It suggests some potential connection, but the details of such a scenario still have to be worked out. It is one more of the many unsolved cosmic mysteries.

The Unsuspected Flip-Flops and Other Astronomical Anomalies

Why do some planet's satellites, including Halley's Comet, revolve in the wrong direction (retrograde)? Why do the innermost planets, Mercury and Venus, have such extremely slow axial rotations? How is it that Venus rotates not only slowly but in the reverse direction relative to other planets? Well, here is a plausible solution: One important clue is provided by our own planet, the Earth.

The ocean floor displays magma which allows us to recognize multiple successive reversals of the Earth's magnetic field. What is this evidence telling us and how are magnetic fields generated? Scientists in the 19th century had already realized that magnetic fields can be produced by circulating electric currents. The Earth has a solid, spinning core which carries a net electric charge.

By changing the direction of the net current flow, the resulting magnetic field will reverse in direction. How does the Earth's core perform this feat? The title of this section provides the answer. The core simply flips over 180°. Multiple core flip-flops thus provided the successive signatures on the ocean floor for the magnetic field reversals. Now let's examine the reason for the core's flip-flop. The success of the flip-flop phenomenon is dependent on two prior conditions:

a. The development of a relatively cooler solid mantle.
b. An independent, rotational motion of the spinning inner core, with a minimum amount of coupling to its outer mantle.

As hot magma was pushed up from the Earth's interior, it must have destabilized the Earth's core such that it required a flip-flop to retain its stability. Notice that field reversals were accompanied by magma ejections, which if arisen directly from the core would have produced a torque (a turning moment) which would have destabilized it. The intervening high-pressure gas between the core and the outer mantle provides adequate decoupling between these independently revolving structures. Hence a large number of flip-flops were able to take place.

Now for the planet Venus, a similar phenomenon was presumably also in progress. It also had a cooler outer mantle and a spinning inner core. During the progression of successive flip-flops, intervening gases eventually dissipated, and (coupling) contact between the core and the mantle presumably took place.

The inner core rotation was opposite to that of the mantle when the final contact coupling phenomenon occurred. This event had cataclysmic consequences: It virtually cancelled out the rotation rate of Venus (rotational momentum cancellation), leaving it in an exceptionally slow reverse rotation. The induced heat generated due to the angular momentum cancellation was expanded predominately in the mantle.

This in turn caused the evaporation of all the water content of the planet's oceans as well as vaporization of the hydrocarbons that were associated with the upper layers of Venus' mantle. The intense solar radiation then caused the dissociation of the water molecules into hydrogen and oxygen, with the hydrogen escaping Venus' gravitational field. Next, the oxygen combined with the hydrocarbon vapors to form carbon dioxide (CO_2) and a small amount of carbon monoxide (CO). This is the state of affairs which is observed on Venus today.

We have so far primarily discussed the flip-flops of the interior cores of planets; however, one must also consider the possibility of outer mantle flipping. Examples of such situations are demonstrated by each of the major planets during their early development stages. Each of the major planets launched one or more satellites while in this "upside down" outer mantle configuration. It is also significant to note that the early satellites launched by these major planets resulted in the retrograde motions of each of these satellites at fairly large distances from each of their parent planets, as well as somewhat significant elongated orbits (with the exception of Triton, whose orbit is almost circular).

The Saga of the Planet Neptune

All planets of our Solar System obey the Bode-Titius law (discussed in chapter 3) with the exception of Neptune. What happened here? Neptune should have occupied a distance of about 38 A.U. instead; its mean orbital distance from the Sun is presently at 30 A.U. How is it that a minor planet, Pluto, occupies the orbit which should have belonged to Neptune?

There are a number of important clues which could help us in solving this Neptune puzzle:

1. Most of the planets have one or more moons "attached" to them. Mercury and Venus are the only lonely planets devoid of moons.
2. Proto-Mars, while in its very early development stage, probably ejected its two small Moons (Deimos and Phobos) from its giant volcano "Olympus Mons".
3. We may therefore extrapolate that some of the satellites of the four major giant planets probably originated in roughly similar ejection processes.
4. Neptune was the firstborn planet. Each planet was extremely hot and contained a large component of rotational angular momentum. Multiple, high velocity ejection processes of molten lava helped to stabilize each of the planets. Since each of the planets was spinning, the ejection velocity always contained both a radial and tangential component.

Since planetary mass ejections occurred at random intervals and at various angular degrees during a planet's complete rotation, their net effect on the planet's orbital velocity is relatively small, with the exception of Neptune. A bulk mass ejection occurred during Neptune's very early existence. It was essentially equivalent to the well-known effect of firing a retrorocket for NASA's space explorations. This process increased Neptune's speed from about 4.8 km/s to 5.478 km/s. It's orbital distance was thus decreased from about 38 to 30 A .U. Hence one may also deduce that Neptune's original mass was substantially greater than it is at the present time.

Conclusion: Pluto plus its companion Charon happened to be minor remnants of the total ejected matter from Neptune. Its net velocity component was approximately parallel to Neptune's original orbit. The bulk of the rest of Neptune's ejecta was flung out in all directions (due to Neptune's spin rate) and now constitutes part of the so-called "Kuiper belt".

As a consequence of Neptune's rotation, Pluto was probably ejected relative to Neptune's surface, with a small tangential component. This could have resulted in Pluto's elongated orbit ($e = 0.2488$), which is by far the largest eccentric orbit of all the planets. Pluto's orbit is in fact out of the ecliptic, so only a portion of its ejection velocity would have contributed to Neptune's orbit's distance change. It is not a 2-D, but a 3-D problem.

Of all the retrograde satellites associated with the major planets, Triton is indeed an exception relative to the other satellites which have rather elongated orbits. Furthermore, they are all at very large distances from their respective originating planets. Triton's large mass and radius differentiate it from the other retrograde satellites, which are orders of magnitude smaller in size and less massive. Furthermore, Triton was spun-off, rather than forcibly (violently) ejected, hence it retains its exact circular orbit.

We have so far primarily discussed the flip-flops of the interior cores of planets; however, one must also consider the possibility of outer mantle flipping. Examples of such situations are demonstrated by each of the major planets during their early development stages. Each of the major planets launched one or more satellites while in this "upside down" outer mantle configuration. It is also significant to note that the early satellites launched by these major planets resulted in the retrograde motions of each of these satellites at fairly large distances from each of their parent planets, as well as somewhat significant elongated orbits (with the exception of Triton, whose orbit is almost circular).

The Johnson Hypothesis for the Origin of the Solar System (continued)

The model that will be considered predates the analysis of chapter 3. Is included here since it provides another perspective and a slightly different type of analysis. In addition, it has features that fill in some of the missing gaps. It again assumes that the original Sun developed from a compact, highly dense, massive, neutron core, whose mass was ninety-six times the present solar mass. It will be shown that this initial mass can also be deduced based on all planets' relative positions in their orbits (Conservation of Linear Momentum principle), plus from a mass-luminosity graph (see Fig. 1, Chapter 1).

In the previous chapter, the solar system's evolution was described. Portions of its early development will now be re-explored by a different method. In particular, the method by which each planet moved from its initial 0.4 AU orbit to successive more remote orbits will be described.

The youngest and hottest stars (such as O-type and Wolf-Rayed stars) are the most luminous and rotate the fastest (as exhibited by their broadened spectral lines). These stars represent the early stages in stellar development. The initial Proto-Sun was also spinning very rapidly. (The rapid outside surface rotation of an O-type stars is a strong indication of an inner-neutron core spinning at an enormously high speed.)

The rapid spinning of the solar core created an extreme instability in the Proto-Sun. Its core splits into two equal components, rotating around one another. This in turn produced a very strong tidal effect: a differential "pulling" force capable

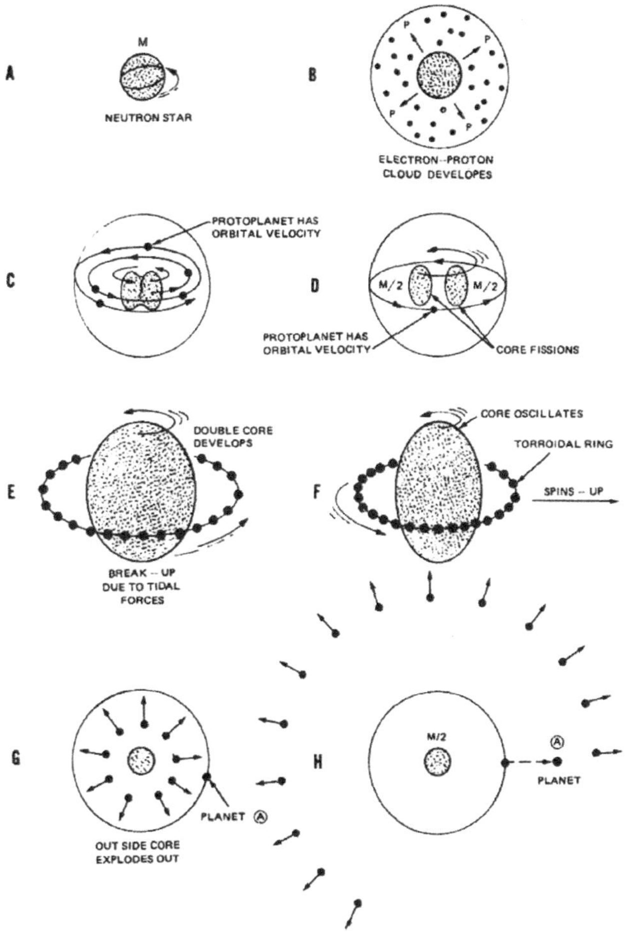

Figure 4.1 *This idealized schematic diagram shows the steps involved in the launching of a planet, and the evolution of the Proto-Sun. (A) Neutron star, compact and spinning rapidly. (B) The evolution of a gas cloud consisting of electrons and protons. Its overall extent brings it to the size of the orbit of Mercury, 0.4 A. U. (C) Development of a central core. A planet achieves orbital velocity at the distance of 0.4 AU. Meantime, the core develops in a series of steps, E, and F The core fissions into two parts, one part (will continue to break up into smaller and smaller components; perhaps billions of them, revolve very rapidly around the remaining core. They achieve enormous speeds eventually, via successive fission processes. All are ejected suddenly, presumably because of their enormous speeds (their effective mass could increase if they had achieved relativistic speeds). In any event, an explosion occurs (nova) with all of them leaving the region of the star. as they pass the proto planet, on their way out into space. As soon as they do, the proto-planet's orbit doubles. i.e., planet A is now held in by half the previous gravitational attraction. In view of the conservation of angular momentum, it will double its previous distance corresponding to the mass loss of the core (Fig. 4.1 H).*

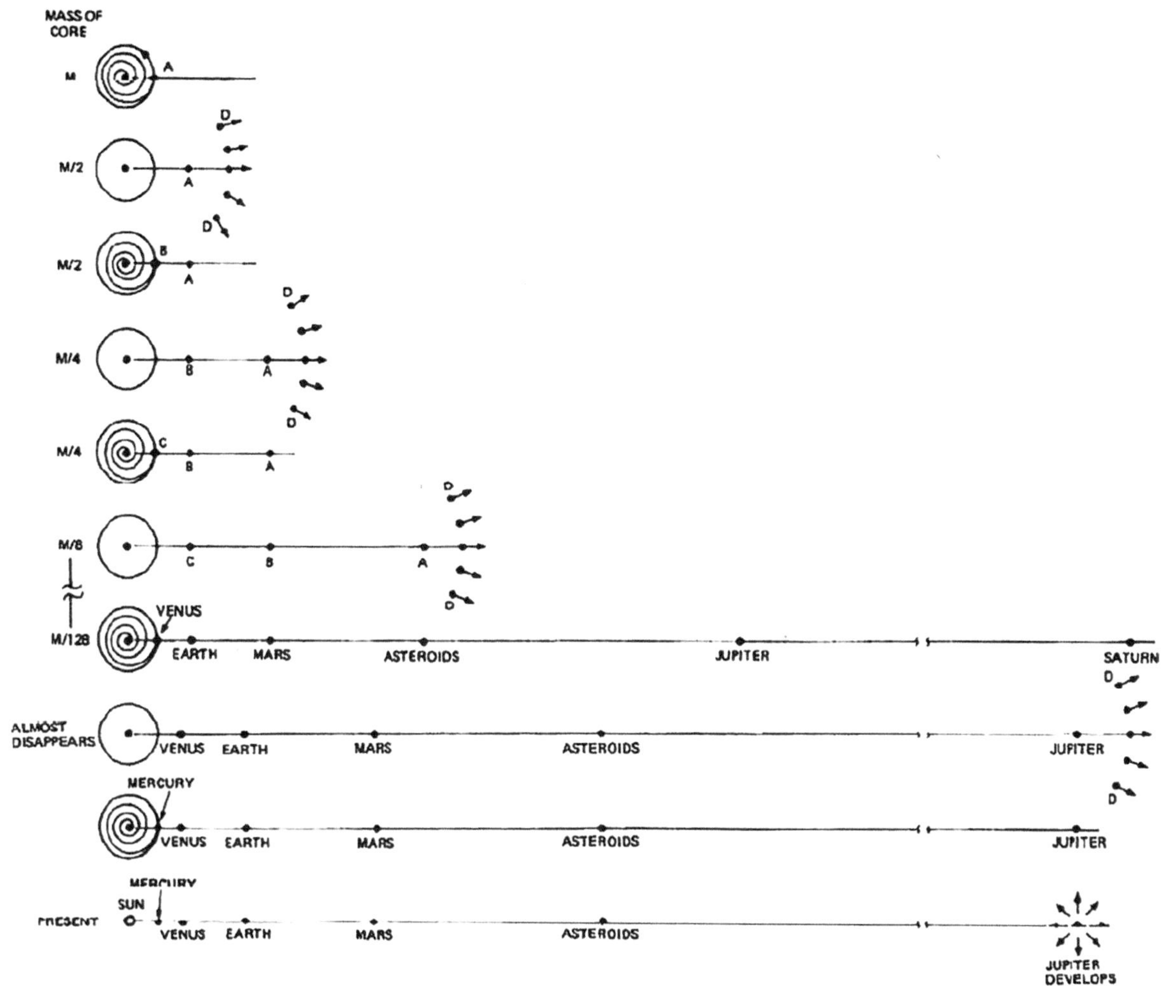

Figure 4.2 *The development of the solar system. Successive stages in the NOVA like particle ejections and the associated launchings of planets into appropriate orbits is shown schematically. The various steps involved are discussed in the text.*

of disrupting its stability. It resulted in multiple fragmentations of one of the cores, viz. millions of components circumnavigating the surviving core in a rapidly revolving ring-like pattern. Figure 4.1 shows possible stages in the development of the core. In chapter 3, it was shown how each planet was produced in an identical orbit at 0.4 A.U. Here is shown a possible scenario for successive halving of the core's mass and the resulting movements of each of the planets.

1. The solar system development (Johnson 2015) originated as a dramatic, successive, explosive production phase, which involved the sequential expulsion of about 95 solar masses, which was part of the successive planetary launch procedure. Most of the explosively-distributed material from the Proto-Sun ended up in the interstellar medium.

Next, a rather dramatic development occurs, when the inner core's revolving particles can no longer maintain stability in their orbits. They were violently ejected, presumably by a Coulomb repulsion mechanism (Nova). Their ejection speeds were large enough to symmetrically escape the gravitational field of the Sun. This effectively halved the mass of the Proto-Sun, and consequently also halved the strength of its gravitational field, as well as the angular momentum of a proto-planet at the Sun's surface. As soon as this "***Collection***" passed the outer surface of the Proto-Sun, the proto-planet doubled its orbital distance from the Sun's center due to the diminished gravitational field.

(This extremely oversimplified development sequence is a broad generalization of what might have happened. However, the details were most likely far more complex, involving a hybrid system of nuclear, gravitational, and complex electromagnetic effects, all interacting at different locations throughout the Proto-Sun, at various development stages.)

So far, the discussion demonstrated the launching of one typical planet. Figure 4.2 shows the complete sequence of planet launchings from beginning to end.

This is followed by successive division of the stellar core and its subsequent explosive expulsions, in a process very similar to the one described for the first expulsion. Each nova explosion comprises the leaving of millions of particles from the Sun. As they all passed the orbital position of each planet, the distances from the core increased in proportion to the change in Proto-Sun mass. At each launch, the Proto-Sun maintains its initial giant radius at 0.4 A.U.

A planet's orbit is governed by the total stellar mass ***within*** the planet's orbit. Any mass ***outside*** a planet's orbit no longer gravitationally affects that planet and can be ignored.

Orbital planes of the planets could have been easily altered as a result of an asymmetric explosion. Except for Pluto, all planet's orbits are indeed almost in the same plane. The fact that Mercury's orbit and the Sun's equatorial plane are coplanar, provides support for the "Johnson Hypothesis". (They are both inclined at approximately 7 degrees with respect to the plane of the ecliptic.)

We shall now derive the Bode-Titius law which was on the books for 250 years. In fact, it was recently denied its official "law" status. The fallacious argument for denigrating it to a lower status was predicting mean planetary distances is just a coincidence of random numbers. Feeble rebuttal voices by some who argued that this so-called law at least does indeed predict the distances quite well were drowned out by shouts of:

1. "It's all just accidental coincidences!" (Eight of them!!)
2. "Any half-baked theory will come up with a logarithm-type sequence." (But not this exact one!)
3. "It does not explain Neptune and Pluto." (That's because the law was not adequately used—see Table 4.1. It works well for Pluto, if you skip Neptune.)

The basic physics behind this law is the conservation of angular momentum. Each protoplanet in orbit at the Proto-Sun surface is assumed to have achieved orbital velocity appropriate to the proto-stellar mass at this stage of its development.

We shall now develop a set of simple equations which contain only two unknowns: (1) The initial mass of the Proto-Sun's core, M_0, and (2) the radius of the orbit from which all the planets were launched, a_0. For simplicity, we shall assume that the size of the Proto-Sun did not change subsequent to each nova explosion. Using this simplified analysis, we can write down eight simultaneous equations with the two unknowns, which are as follows (using the fact that the expression for the angular momentum is given by Equation 1):

To see why the distance would be doubled, consider the equation below:

$$J = m \cdot \sqrt{G\left(M_0 + m\right) \cdot d \cdot \left(I - e^2\right)} \quad \ldots\ldots\ldots\ldots\ldots\ldots (1)$$

(Equation 8.30, Motz & Duveen)

This equation gives the angular momentum of each of the planets. Note that it contains the gravitational constant (G), the orbital distance (d), the mass of the Proto-Sun core (M_0), and the mass of the orbiting planet (m). Since the mass of the orbiting planet is much smaller than the core's mass, one can ignore it for present purposes. For simplicity, the eccentricity factor (e) will also be ignored, since circular motion is assumed, rendering its value close to zero. Using these simplifying assumptions, Equation 1 reduces to the following:

$$J = \left(constant\right) \cdot \sqrt{\left(M_0 \cdot d\right)} = Constant \text{ for any proto - planet's angular momentum} \quad \ldots\ldots\ldots (2)$$

Moreover, its energy (E) is also a constant:

$$E = -\frac{GMm}{2d} \quad \ldots\ldots\ldots\ldots\ldots\ldots (3)$$

It now becomes apparent based on Equations 2 and 3, that as soon as the mass of the Sun's core is halved, in order to maintain the same angular momentum and energy, the distance d must be doubled. The physical principles are: conservation of energy and angular momentum. Since both the angular momentum and the energy of each planet must be conserved, one can readily see that by successively changing the mass of the Proto-Sun's core, the distance of each planet will increase in stages, which are proportional to the change in the Proto-Sun's mass.

Instead of writing all eight equations, we shall only write three of them, which will illustrate the pattern of the equations. Note that after each nova, the increase in distance corresponds to the ratio of:

$$\frac{\text{solar mass before nova}}{\text{solar mass after nova}}$$

For a succession of such novas, cancellation of factors in the numerator and denominator must occur, resulting in a set of very simple equations.

Let:

M_{Sun} = mass of original core = M_0

a_0 = orbit at launch

We assume that every proto-planet, in turn, has achieved orbital velocity at the Proto-Sun's surface (radius a_0) prior to its launch via nova explosion.

For example, for Mars:

$$\frac{\left(1+\frac{M_0}{32}\right)}{\left(1+\frac{M_0}{64}\right)}\cdot\frac{\left(1+\frac{M_0}{64}\right)}{\left(1+\frac{M_0}{128}\right)}\cdot\frac{\left(1+\frac{M_0}{128}\right)}{(1+0)}\cdot a = 1.523\ \text{AU} \quad \text{.......... (4)}$$

Where one AU is the mean distance between the Earth and the Sun, approximately 93 million miles.

Also note the cancellation of factors in numerator and denominator.

For Earth:

When the Earth was launched from the Proto-Sun surface (following solar nova No. 7), its distance increased from a_0 to:

$$\frac{\left(1+\frac{M_0}{64}\right)\cdot a_0}{\left(1+\frac{M_0}{128}\right)} = \frac{\left(1+\frac{6}{4}\right)\cdot a_0}{\left(1+\frac{3}{4}\right)} = \frac{10}{7}\cdot a_0 \quad \text{.................. (5)}$$

$$\left(1+\frac{M_0}{64}\right)\cdot a_0 = 1\ \text{AU} \quad \text{.. (6)}$$

For Venus:

By the time proto-Venus spiraled to the Proto-Sun's surface, reaching orbital velocity at a distance of 0.4 AU, the remaining core mass of the Proto-Sun was only 1/128 of its original mass M_0.

At this point, we shall postulate that one of the remaining double cores disappeared. Incidentally, there is never an exact doubling of distance because of the (M/n+1) factor, $\left(1+\left(\frac{M_O}{128}\right)\right)$ for the case of Venus.

The "1 factor" is proportional to the remaining mass of the Sun, including its gaseous envelope. Each planetary distance subsequent to Venus' Nova increased in proportion of $\left(1+\left(\frac{M_O}{128}\right)\right)$ to 1, which corresponds to 1.75.

$$\left(1+\frac{M_0}{128}\right)\cdot a_0 = 0.723\ \text{AU} \quad \text{.................................... (7)}$$

Finally, when the last remaining proto-planet, Mercury, was on Proto-Sun surface, there were no more nova explosions. Mercury launched itself at the Proto-Sun surface and remained there for all time in the orbit it is now in.

There may be several possible mechanisms for the Proto-Sun to have maintained this large initial orbit and to have shrunk, (most likely involving electrostatic interactions.) i.e. The last double core configuration revolving about their common center of gravity both were negatively charged. By halving this charge, a new equilibrium distribution was established.

The right-hand side of Equation 7 is the present distance of the planet Venus.

All eight equations, with only two unknowns, are consistent with $M_0 = 96\ M_\odot$ and $a_0 = 0.393$ A.U. Note that this initial solar mass $M_\odot$ is also consistent with the graph in Chapter 1 where the stars' luminosities are plotted against their masses, and where the origin (where lines A and B meet in Figure 1) corresponds to the large initial mass $M_\odot$ of the Proto-Sun.

At Neptune's and Uranus' launch, the initial change in the Proto-Sun's core mass $M_\odot$ is much larger compared to the present solar mass. Thus, neglecting the present solar mass in these equations did not affect the final solutions. However, when considering the last of the nova events, when the Proto-Sun's remaining mass is closer in value to the current Solar mass, its value does become significant and must be included in the equation.

Note that when the Earth is launched, it increases its distance by a factor of (10/7). It does not actually double its distance from the Proto-Sun radius, because the factor of one is already significantly large in comparison with the remnant solar core. Its fractional increase in distance is an additional 7/4 when Venus is finally launched. At the time of Venus' nova, it is assumed that one of the two stellar cores disappears via another nova-type expulsion.

This calculation produced eight over-determined equations with only two unknowns, namely the initial radius of the planets' orbits and the initial Proto-Sun's core mass. Consequently, in principle, one could readily improve on the results, as was done for Jupiter.

The Bode-Titius Law of 1772, Interpreted by F.M. Johnson in 1972:

$d = 0.4 + (0.3) \cdot 2^{8-n}$ where n is an integer = order of birth.

Mercury ($n = \infty$) signifies that it was left in the original orbit from whence all its sister planets originated. No core fission or explosion was necessary to launch Mercury. Subsequently, the Proto-Sun gas cloud shrank to its present size.

Table 4.1 is a summary of the calculated distances and the actual planetary mean distances. This first calculation

Table 4.1

	Assumed Protoplanet Mass (m)	Actual Distance	Calculated Distance	% Error
Pluto		39.44	38.1	+3.3
Uranus	Not Yet Used in this Calculation	19.13	19.2	0
Saturn	Not Yet Used in this Calculation	9.54	9.8	–2.7
Jupiter	0.2 $M_{\odot}$	5.20	5.2	0
Ceres		2.77	2.73	0.6
Mars		1.523	1.52	0.2
Earth		1.00	1.00	0
Venus		0.723	0.73	–1
Mercury		0.387	0.393	–1.5

had assumed that a_0 did not change and neglected the mass of the proto-Jupiter in the angular momentum equation. The improvement becomes apparent if one takes into account the mass of Jupiter (about 0.2 M_0) upon launching. The results, when this is done, are also indicated in Table 4.1. There is satisfactory agreement with the actual planetary distances.

The cause for the large number of meteor craters in planets throughout the solar system now becomes apparent. While Mars and the Earth have been substantially altered by weathering effects, obliterating some of the meteor impacts, this is not the case for the Moon or Mercury.

Jupiter, Saturn, Neptune, and Uranus require further discussion. Jupiter, prior to its evolution, actually had sufficient mass to be characterized as a star. Its initial mass was probably about 0.2 solar masses, as described previously. The solar system, at that point, was in fact a multiple-star system.

Some of the relics of our early solar system evolution are still discernible at the present time: they include the orbital eccentricity of the planets.

They are indicative of the above analogous solar system evolution in the galaxy. (Planets are being generated!) A Supernova, on the other hand, allows us a rare glimpse into a possible unsuccessful (failed) solar system development. A supernova thus permits a direct view of a star's interior, since its plasma envelope and subsurface structure are ejected while its remaining neutron core is thereby exposed. The related pulsar e-m emission signals arise from orbiting SNFRs in the presence of a strong nuclear core magnetic field.

The Earth Moon System

In the distant past, the Earth's diameter was much smaller than its present value. The implication of this intermediate result is that the Earth's internal average density was at least eight times larger at an earlier epoch in the Earth's history. Since its present *average* density is 5.5 gm/cc, the implied earlier *average* density of 44 is much larger than any known density of metals, even under pressure. In fact, since the present density of the Earth's *central core* is approximately 13 an even greater ancient *central core* density can be extrapolated (over 80). With this background information, a further analysis will require consideration of the Earth-Moon system.

Since rotation of the Earth and revolution of the Moon occur about a common center of gravity, we can think of them together as the "Earth-Moon system". Because this is one entity, the law of conservation of angular momentum (L) can be invoked to describe the system.

When the present angular momentum of the Moon is added to that of the Earth, the law of conservation of angular momentum guarantees that the calculated total momentum of the system was the same some 4.5 billion years ago.

At that time, the Earth was spinning exceedingly rapidly, and needed to be stabilized. If our present Earth, which is rotating at a speed of about 1000 mph at the equator, were to increase its speed to 24,000 mph, anything at the equator would leave the Earth for outer space.

The same concept is proposed for the launching of the Proto-Moon from the Proto-Earth. By means of the spin-off of a portion of the outer, molten surface of the young

Proto-Earth, the beginning of a necessary stabilization was achieved. The results were as follows (the details of this analysis are exhibited starting with Equation 8):

Following the Proto-Moon's release, the Proto-Earth had a radius of 357 km and a rotation period of 48 seconds, which is a factor of about 1800 faster than its present 23 hr, 56 minute rotation period. In addition, the lunar tidal effect that resulted from the Proto-Moon's revolution about the Proto-Earth provided tidal friction, which gradually allowed the Earth's spin to slow down. The Proto-Earth's gradual diameter expansion also contributed to this slowing. This "spin down", when averaged over 4.5 billion years, corresponds to a slow-down of roughly two milliseconds per century, which is confirmed by observation.

Units: The c.g.s. system is used for this analysis:

The moment of inertia (I) of the solid Earth is given as

$$I = \frac{1}{3} M_E a^2 \qquad (8)$$

Where M_E = mass of Earth

a = mean Earth radius

The angular momentum of the Earth (L_E) is given by

$$L_E = I_\omega \qquad (9)$$

where ω = angular rotational velocity (radians/sec)

The handbook values (Cox 2000) for the Earth are:

$L_E = 5.861 \times 10^{40}$ cm^2 g sec^{-1}

$\omega = 7.292 \times 10^{-5}$ rad sec^{-1}

The Moon's orbital angular momentum (L_m) is equal to the product of its mass (m), orbital velocity (v) and mean distance (r) from the Earth. As a first order calculation:

Thus, $L_{moon} = m \cdot r \cdot v$

Where, $m = 7.35 \times 10^{25}$ gm

$r = 3.844 \times 10^{10}$ cm

$v = 2\pi r/P$

where P = Moon's sidereal period = 236×10^6 seconds = (27.32 days)

Substituting these values:

$L_{moon} = 2.89 \times 10^{41}$ cm^2 gm sec^{-1}

$L_{Earth} = I\omega = 5.86 \times 10^{40}$ cm^2 gm sec^{-1}

$L_{moon} + L_{Earth} = 3.48 \times 10^{41}$ cm^2 gm sec^{-1} = constant ...... (10)

The escape velocity from the Earth is $V_e = \left(\frac{2GM_E}{R_E}\right)^{1/2}$

Where, G = gravitational constant

M_E = Earth Mass

R_E = Radius of Earth at the time of spinoff

This equation can be rewritten in terms of angular velocity ω_e, since $V = \omega_e \cdot r$

$$\omega_e = \left(\frac{2GM_E}{(R_E)^3}\right)^{\frac{1}{2}} \qquad (11)$$

The connections between rotational period P and angular velocity ω is

$$\omega_e = \frac{2\pi}{P} \qquad (12)$$

Solving for the angular velocity ω_e, when the Moon of mass m fissions off the Earth:

From Equation 10,

$L_{moon} + L_{earth} = 3.48 \times 10^{41}\, cm^2\, gm\, sec^{-1}$

$$m\omega_e (r_e)^2 + \frac{1}{3} M (r_e)^2 \omega_e = 3.48 \times 10^{41} \qquad (13)$$

$$(r_e)^2 \omega_e \left[m + \frac{M}{3} \right] = 3.48 \times 10^{41}\, cm^2\, gm sec^{-1} \qquad (14)$$

substituting for ω_e from Equation 11, and putting in numerical values, one gets $r_e = 0.357 \times 10^8\, cm =$ **357 km**

(Note: present solid Earth core radius is 1221.5 km, only a factor of 3.42 times larger than the Earth's original radius)

Now solving for $\omega_e = 0.132$ radians/sec.

And $P_0 = \frac{2\pi}{\omega_e} = 47.5$ seconds

In summary: Earth's radius increased by 17.86 times

$$\omega_e / \omega = \frac{1.32 \times 10^{-1}}{7.29 \times 10^{-5}} = 1810 \text{ (spin down)}$$

Thus, the Earth's volume initially was $(17.86)^3$ smaller i.e. 5,700 times greater density.

Today, average Earth density is 5.5 gm cm^{-3}. Hence, the average Earth density at Moon's spin-off was about 31,300 gm cm^{-3}

This enormous early Earth density confirms its solar origin and its nuclear fission energy source.

The Neptune Triton System

Could another planet in our solar system have undergone a similar development as that of the Earth-Moon system? Indeed, it appears that Neptune, the first major planet to be launched, had a similar experience. Its largest Moon, Triton, has a mass of 2.14×10^{25} grams, whereas Neptune's mass is 1.02×10^{29} gm. Due to its tidal effect, Triton's rotation rate is synchronous with its revolution about Neptune, this is analogous to the Earth-Moon system.

Table 4.2

	Earth	Neptune	Moon	Triton
Mass	5.976×10^{27} g	1.0244×10^{29} g	7.348×10^{25} g	$2.15\text{x } 10^{25}$ g
Radius	6.371×10^{8} cm	2.476×10^{9} cm	1.737×10^{8} cm	1.352×10^{8} cm
Average Density Now	5.5	1.76 (core 3.54)	3.3341 gm/cc	2.07 gm/cc
Birth Density	3.13×10^{4} gm/cc	7.2×10^{5} gm/cc		
Rotation Rate ω_e	0.132 radians/sec	0.60 radians/sec		
Initial Period	47 sec	10.5 sec		
Original Radius	3.57×10^{7} cm	3.36×10^{7} cm		

Triton's period about Neptune is 5.8768 days and its mean distance from

Neptune is 3.5476×10^{10} cm.

Thus, Triton's orbital angular momentum $L=m\cdot v\cdot r=$ 3.33×10^{41} c.g.s.

Neptune's rotational angular momentum $=1.97\times10^{43}$ c.g.s.

Using the same type of calculation as was done for the Earth's Moon, the following results are obtained:

$\omega_e=0.60$ radians/ sec and $r_0=3.36\times10^{7}$ cm at spin off,

Neptune's "original" rotation period was 10.5 seconds

Since Neptune's radius is now 2.476×10^{9} cm, its size has increased by a factor of 74.2 times and its volume has increased by $(74.2)^3=408{,}000$ times since its initial launching. This implies that Neptune's original average density was that much greater. It is 1.76 now and hence was originally $\rho=7.2\times10^{5}$ gm/cc

The retrograde motion of Triton implies that Neptune's spin axis had flipped 180 degrees, at some point in the past, prior to Triton's launch!

Compare the original Earth and Neptune in their post solar birth state (Table 4.2).

Note that the Earth and Neptune had very comparable sizes at birth. However, their original densities and rotation rate were quite different. Neptune's original density was (7.2×10^{5} gm/cc), about 23 greater than Proto-Earth's, and its rotation rate was (0.60/0.132), about 4.5 times greater than that of the Proto-Earth.

This is an important result since it gives us a clue regarding the Sun's core development, as it sequentially gave birth to its planetary off springs. The large initial planetary densities provide a clue to the Sun's original core density, which had diminished substantially by the time its sixth offspring (the Earth) was launched. Similarly, as might be expected, the Sun's rotation rate diminished subsequent to each planetary birth.

Caveat: There is one significant difference between the Earth-Moon and the Neptune-Triton systems: Neptune had a large number of satellite launchings (in addition to Triton). Thus, for Neptune, we presently do not have an accurate value for its ancient TOTAL angular momentum. Consequently, the quoted results for Neptune/Triton should be considered as preliminary estimates only, and not as a final result.

Sun vs. Planets

The Sun's present mass is 1.989×10^{33} gm. Its present rotational angular momentum is merely ($L=I\omega$) 1.63×10^{48} cm^2 gm sec^{-1} based on its surface rotation (Allen P. 340). Whereas, the total angular momentum of all the planets is 3.148×10^{50} cm^2 gm sec^{-1} (Allen P. 293) i.e. it is 193 times larger. **This single indisputable fact is a major dilemma for present solar system theories.** How did all this angular momentum get transferred to the planets (mainly Jupiter, Saturn, Uranus and Neptune)? The theory below gives the answer:

Assume that the Proto-Sun was a 96 solar mass neutron star, whose mass was 1.92×10^{35} grams. This large initial solar mass is consistent with the intercept shown in Figure 1, Chapter 1, plus the calculated value based on the planetary distance calculations based on the previous derived previously. Nuclear density is given by Fermi ("Nuclear Physics" p. 6) as 10^{14} gm/cc. This is based on his formula $R=1.5\times10^{-13}A^{1/3}$ (from the liquid drop model and alpha decay), where R is the radius of the nucleus, and A is the atomic number. By substituting appropriate values for the U-235 atom, a density of 1.17×10^{14} grams is obtained using this formula.

The volume of this solar neutron star is equal to

$$\frac{Mass}{Density}=\frac{1.92\times10^{35}}{1.17\times10^{14}}=\frac{4}{3}\pi R_0^3$$

Thus, $R_0 = 0.72 \times 10^7$ cm .... (15)

Next, the angular momentum (L_0) of the Proto-Sun is derived using this value for its radius.

$$L_0 = I \cdot \omega_0 \quad (16)$$

Where I = moment of inertia
ω_0 = angular radial velocity (radians/sec)

For a solid sphere, $I = \frac{1}{3}(mass)\cdot(radius)^2$ .... (17)

Substituting values for R_0, and equating it to the total angular momentum of the solar system:

$$L_0 = \frac{1}{3}(96)\cdot(2\times10^{33})\cdot R_0^2\cdot\omega_0 = 3.148\times10^{50} \text{ c.g.s.} \quad (18)$$

Solving for $\omega_0 = 158$ *radians/sec*

Now, the surface velocity V of the nuclear Proto-Sun is given by

$$V = \omega_0 \cdot R_0 = (158)\times0.72\times10^7$$

Therefore, $V = 1.13\times10^9$ *cm/sec*

This is not yet the final result, since an additional amount of angular momentum was removed from the Proto-Sun during its 95 solar mass ejection phase. Thus, the derived value for V should be considered a minimum value (V = 3.7% of the speed of light). The final V value in any event could never be greater than c (but could possibly be an order of magnitude greater than the derived value of 3.7% c).

Please note that all of the above detailed analyses confirm our big picture of galactic and solar system development scenarios in (Chapter 11).

(This analysis is consistent with the Second Law of Thermodynamics).

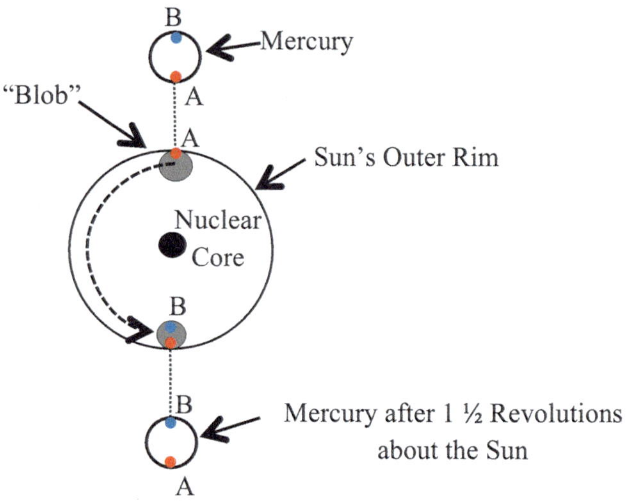

***Figure 4.3** illustrates the initial position A of Mercury and a tight conglomeration of SNFRs. The next critical opposition event, when a tidal effect would again be transmitted, would occur at position B, after Mercury had revolved around the Sun 1 ½ times. In this particular configuration, the tidal effect between the SNFR A and Mercury's surface would occur on the opposing side of Mercury. Assuming this process occurred a few dozen times, sufficient tidal and repulsive forces would have manifested themselves on Mercury, such that its orbit would have achieved its present elliptical shape. This is also how Mercury's pronounced 3:2 tidal lock-in configuration was simultaneously produced. This effect is still visible to this day, even though its initial origin is no longer operating. (This peculiar effect is now explicable using the Johnson Hypothesis.)*

**Not drawn to scale*

Appendix (A Tentative Theory)

Mercury's Unexplained Elliptic Orbit and its 3/2 Tidal Effect (A new hypothesis)

The following is the most probable scenario for Mercury's launch (with many details still to be worked out).

The Sun once operated as a supergiant star. At the time when Venus was on the Proto-Sun's surface ready for launch, the Proto-Sun's core consisted of two identical core components spinning about a common center of mass.

An event that produced a negative surface charge on each of the rotating cores resulted in a compromised equilibrium condition within the Proto-Sun. A resulting tidal effect then broke up one of the cores into smaller pieces, which then left the Sun's surface in a nova event.

With only half the repulsive electrostatic charge now in operation, the Proto-Sun's radius then collapsed to the present size of the Sun. As a result, the fission reactor components, which had previously been on the surface of the Proto-Sun, now were forcibly thrown together as they retreated to their much smaller orbital radius. This in effect produced a molten conglomerate, which made up the protoplanet Mercury. There was therefore a very brief moment in the history of the Sun when its luminous output was essentially zero. (Note the mass of Mercury is exactly equal to the Sun's subsurface mass Q).

In sequential steps, it would then once again build up its gas-cooled fission reactor components by the launching of SNFRs/Sun spots from its neutron core towards the (newly reduced) present-size proto-stellar surface.

With this possible creation narrative for the proto-planet Mercury, it is now possible to interpret some of the peculiar effects and relationships that exist between Mercury and the Proto-Sun. Two of these are examined below.

Mercury displays a very peculiar relationship between its rotation period and its period of revolution about the Sun. Its revolution about the Sun is 87.97 days, and its rotation period is 58.646 days. The ratio between these two values is exactly 1.5. What is this ratio's origin?

Following Mercury's creation in its present orbit, the Sun's surface fission reactor had not yet fully regenerated. However, a few Sun Spots would have been launched, and revolving around the Proto-Sun's core.

Mercury, the solar subsurface SNFRs, and the Proto-Sun's core were all negatively charged, hence their periodic alignment during their revolution in orbit (see Figure 4.3) would have resulted in an extra repulsive force. This force, though small, eventually caused a cumulative effect on the planet Mercury, providing it with a more elliptic orbital shape, as well as introducing a 3:2 tidal effect.

References

Cox, A.N. 2000, Allen's Astrophysical Quantities, Springer-Verlag, New York

Hernqvist, K.G. & Johnson, E.O. 1955 *Phys Rev*. 98, 1576

Chapter 5

The Earth

Synopsis

Credit: NASA/ GSFC/ NOAA/ USGS

This chapter will provide a new outline for the Earth's development during the last 4.5 billion years. It started out as a hot, molten nuclear fission core.

One obviously cannot do justice to all topics related to the Earth in a single chapter. However, those aspects that have a direct bearing on this book's theme will be briefly discussed. These include the chemical composition of the Earth's crust, ages of rocks, and the evolution of the Earth itself as an example of a typical body within our solar system. The Earth, being the planet we live on, provides the most valuable clues at our disposal regarding the origin of our solar system. It contains an enormous amount of valuable data. It should be studied far more extensively than it is at the present time.

Nuclear Time Clocks

We shall briefly discuss the problem of age determinations since the subject has direct astronomical implications. How old is a given rock on the Earth, on the Moon, or a piece of meteorite? Just as a person's age might be estimated by the state of the skin, the lack of hair, or his teeth, a subtle method is used to determine the age of a rock. The time markers within a rock are specific radioactive atoms, which decay. The spontaneous decay of atoms, such as potassium 40, rubidium 87, uranium 238, and thorium 232 are excellent time clocks. One can perform measurements in the laboratory on all these compounds and thereby establish their decay rate. The only assumption one makes in this technique is that nothing within the rock was disturbed since its inception: no daughter or parent products were introduced at any time. Consequently, from the rate of the decay and by knowing the remaining parent-daughter atom ratio within the rock, one can extrapolate back to the age of the parent atoms, or the time their clocks were started.

Let us consider briefly an analogy to illustrate the technique. Imagine that you had deposited $100 in your bank account and that one dollar was removed automatically (as a service charge) from that account every day. Suppose that one day you looked at your bank account and you saw that you had $42 left. Even if you had failed to keep track of the elapsed number of days since the money was deposited, you would immediately be able to deduce that 58 days had elapsed. This analogy illustrates a similar technique used for radioactive atoms.

The half-life of an element is used as a criterion for its decay rate. The definition of a half-life is the time required for one-half of the original atoms to have disintegrated into daughter products. Thus, taking the example of uranium 238, if one starts out with a million such atoms, then, after four and one-half billion years, only half a million atoms of U^{238} will be left. Nature has kindly provided us with a large number of radioactive species to work with. However, only the very long lived ones are of any use in this type of age determination, since one requires some leftover atoms of the original parent in order to establish a meaningful age.

This technique was used to obtain a measurement of the age of our solar system, based on rocks found in meteorites and in a few lunar samples. The oldest rocks found on Earth so far do not exceed 3.3 billion years. (It can be safely assumed, however, that the Earth is 4 1/2 billion years old, based on meteorite and lunar samples measurements.)

Radioactive dating shows that there are certain elements in existence on the Earth today that could not have been there when the Earth was formed 4.5 billion years ago because their half-lives are short relative to this timespan.

The only feasible explanation for the current presence of radioactive species on Earth is the existence of a nuclear fission core within the Earth that operates continuously and

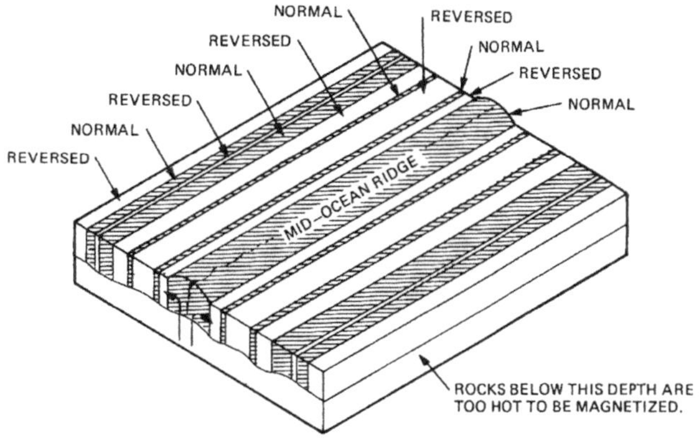

Figure 5.1 *Sea floor spreading and the "frozen" record of the Earth's past magnetic fields. Note how the molten hot material (lava) is pushed up from inside the Earth and how it becomes deposited in successive striped layers as it solidifies. The resulting magnetization pattern is completely symmetric on both sides of the ocean ridge. The record of the Earth's magnetic field present at the time the molten lava is extruded becomes permanently recorded as soon as the lava solidifies.*

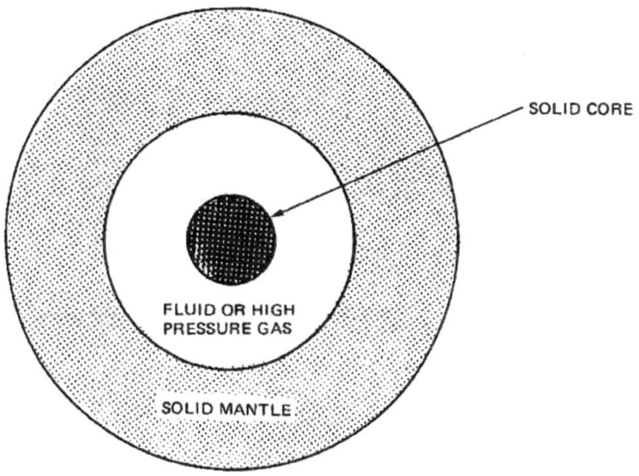

Figure 5.2 *schematic diagram of a cross section of the Earth, showing the solid mantle, a so-called "fluid core" (which far more likely is a gas under high pressure), and the solid innermost nuclear fission core.*

produces such atomic elements. Moreover, its average density is 13 gm/cc, consistent with an assumed 65% average uranium and 35% other compounds content (see Chapt.1).

Geothermal Temperature Gradient

Introduction to the Earth's Structure

The geology of the Earth, its internal structure, and its innermost properties contain, in a large measure, the secrets and the evidence for the early evolution of the solar system. Unfortunately, this evidence is difficult to accumulate, and, as far as the inside of the Earth is concerned, even more difficult to come by.

In fact, the only technique presently available for mapping in a rather crude way the innermost regions of the Earth is the method of seismology. This method uses sound waves of different frequencies which can travel the 8,000 miles from one side of the Earth and back again, if necessary. By studying the speeds of these waves, the manner in which they are reflected, and the details of the two types of waves (namely the shear waves and the pressure waves) one can unambiguously determine whether the waves are travelling through a solid, a liquid, or a gas. Unfortunately however, one cannot distinguish between a liquid and high-pressure gas. The key results are that the Earth contains a solid core surrounded by a fluid or a high-pressure gas core, then by solid mantle (Fig. 5.2). The density distribution inside the Earth is shown in Figure 5.3. In addition, during the last 60 years or so, a very complete mapping of the rift valleys, the trenches, and the "Ring of Fire", as the Earthquake belt is called, has indicated that the Earth is still in a very active mode: the continents are drifting apart.

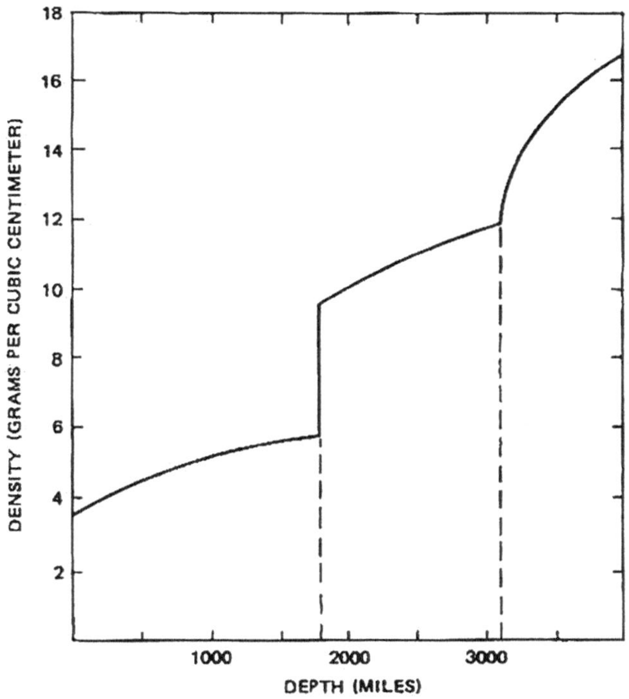

Figure 5.3 *The Earth's density as a function of depth. This data was deduced from seismic measurements.*

The Earth is Hot

Scientists have long realized that the inside of the Earth is hot. The argument is how hot?

The evidence for the increase of temperatures is undisputed as one drills deeper into the Earth's crust. The temperature increases about 1° C for every 100 ft. of depth, or about 50° C per mile. Work by D.I. Turcotte and G. Schubert indicates that at depths of about 300 kilometers, temperatures as high as 1,500° C are encountered. If one extrapolates this rate of temperature increase, the center of the Earth's temperature would be in excess of 200,000° C.

However, such high core temperatures are not accepted by present-day geologists. Present incorrect theory presumes that internal heat is a result of subsurface disintegration of radioactive species. Such theories cannot account for more than a few thousand degrees within the Earth.

These seemingly impossibly high temperatures, however, may actually be a measure of an equivalent high ***potential*** energy source whose energy production capability is ***equivalent*** to these high temperatures. A more realistic source for this heat is a nuclear fission reactor at the center of the Earth, which could have an "equivalent temperature" as large as five million degrees.

As long as energy is extracted at the proper rate, the core temperature stays below the melting point of its containment capsules. Once its energy is no longer removed rapidly enough, its core temperature would rise quickly and cause a complete melt down of the core, as has been observed, for example, in Chernobyl and Fukushima: unfortunate examples of this potential catastrophe. The Earth's core can be approximately interpreted as a large-scale version of a manmade nuclear device.

The Earth's core is gas-cooled by an envelope that surrounds it. When this gas reaches high enough pressures and temperatures, it pushes through fissures in the Earth's mantle, which liquefy it through the Venturi effect producing petroleum and hydrocarbon gasses. Another result is the molten liquid lava, which emerges in volcanoes. In this way, the Earth's fission core is cooled and its excess energy is dissipated.

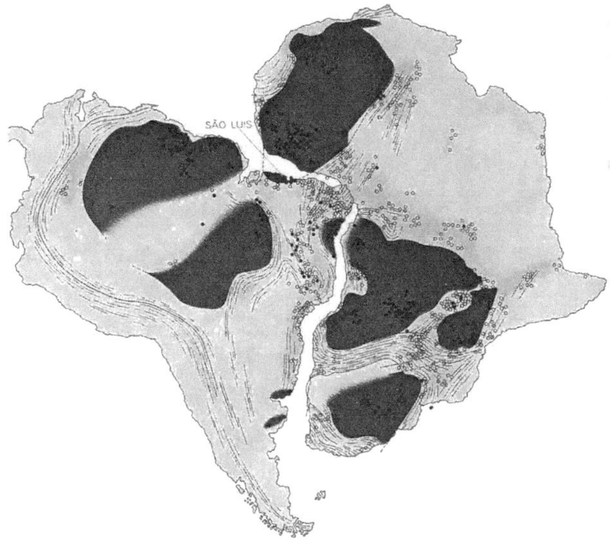

Figure 5.4 *Tentative matching of geological provinces of the same age shows how South America and Africa presumably fitted together some 200 million years ago. Dark areas represent ancient continental blocks, called cratons, that are at least 2,000 million years old. From The Confirmation of Continental Drift by Patrick M. Hurley. Copyright © 1968 by Scientific American, Inc. All rights reserved.*

Figure 5.5 *The Earth now and about 4 billion years ago. Note that all the continents can be fitted with reasonably good precision on a sphere of about 1/2 size of present Earth. Conclusion: The Earth expanded. Implication: A large internal energy source must exist inside the Earth. It is the solid inner nuclear core.*

Earth's Expansion

Geological studies have identified trends in continental movement and oceanic rifting. Using this data, a map has been constructed that shows the supposed initial configuration of the Earth, with all the continents joined together as one mass.

At one time in the distant past, the Earth had approximately half of its present diameter. Evidence of this comes from the fact that all Earth's continents can be fitted together like a jigsaw puzzle on a smaller Proto-Earth globe (Figures 5.5, 5.6, 5.7, 5.8). Corroboration of this theory would involve geologic matching of other parts of the continents which these pictures predict were once joined.

There is geological evidence that 3.5 billion years ago, the *surface temperature* of the Earth was 50° C - far hotter

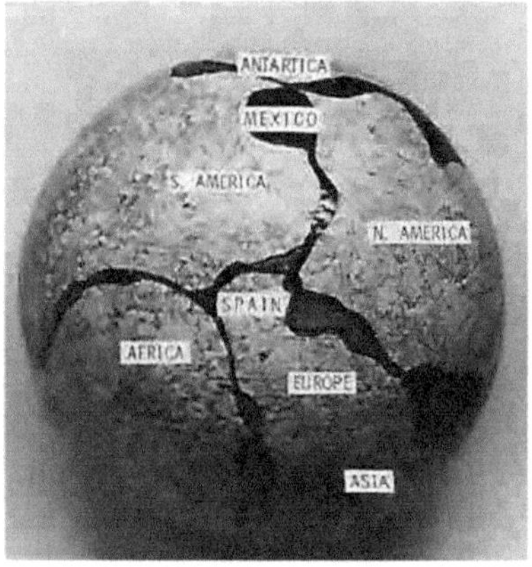

Figure 5.6 *Reconstruction of the continents on globe 1/1.86 present size. Dark areas are overlap or gaps. Note the matching of continental margins on other parts of the continents not previously considered.*

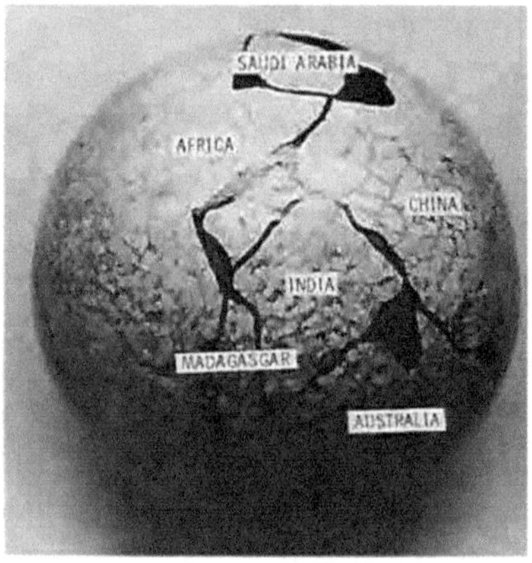

Figure 5.8 *The reconstructed Earth, as seen from different vantage points on the earlier size globe. The orientation of the globe is chosen in order to exhibit all the different continental margins.*

Figure 5.7 *Fitting of continents on earlier size of Earth.*

than it is today. The Proto-Earth then was an extremely hot, molten object whose crust cooled over time. Breaks in the cooling crust allowed molten material from beneath to be pushed up, resulting in the Proto-Earth's expansion.

Continued evidence for this expansion caused by excess energy in the Earth's core can be seen in volcanic activity, mountain-building, hot lava, and infrared radiant energy.

Additional evidence for the Earth's expansion is based on the Earth's slowing rotation rate, since its formation.

Scientists have determined that the Earth's rotation is slowing down at the rate of 1/1,000th of a second per century. If one were to project this slowing rate backwards in time, it becomes clear that 4.5 billion years ago, the Earth must have been rotating much faster.

Evidence for the Earth's slowing down is also contained in some fossil records. Paleontologists have been able to count the daily growth rings in corals and other fossils which suggest that the Earth had about 400 days in the year during the Devonian Period, roughly 375,000,000 years ago.

Magnetic Fields

There are rifts (cracks) in the Earth's crust through which some of the internal pressure, generated by the fission reactor, is released in the form of gas and lava.

These rifts in the ocean floor act as a timeline, since new molten ejected material pushes older material outward. As each new layer of hot lava cooled, the magnetic chemicals within it oriented themselves along the then-existing Earth's magnetic field. It then "froze" in that magnetic field direction, and a permanent record was established. When this timeline record is now studied, it becomes evident that in the distant past, *the direction of the Earth's magnetic field changed many times* (Figure 5.1). The cause of these magnetic reversals is not well understood. However, the Johnson Hypothesis allows its interpretation by two different potential mechanisms:

Charge Reversal

All magnetic fields require moving charges for their creation (basic physics). The hypothesized nuclear fission core possesses a negative charge, which fulfills this requirement.

[By adding electric charges to any spinning system, magnetic fields are generated. If the charges are positive, the magnetic field is in one direction. If the net charges are negative, the magnetic field direction is reversed, even if the spin direction is unchanged.]

Core Flip

It is more likely that the central spinning nuclear core flipped in order to create magnetic field reversals. This is possible since its spin has to be independent of its surroundings. Evidence for its independent spinning was already evident by noting the Earth's magnetic dipole field variation in both magnitude and tilt direction as a function of time. From 1945-1990 its strength diminished by 3% and its orientation changed by about one degree. A flip, however, would require a torque on the Earth's spinning core. This might arise during its occasional material ejection phase, when the ejected material passes through the gas envelope and makes contact with the nuclear solid mantel surface.

References

"Fission products (by element)". *Wikipedia* (2017).

Chapter 6

Life, Energy, and the Interstellar Medium

Synopsis

Some of the material in this chapter can also be found in my textbook *Voyage into Astronomy*, 2nd Edition. It is included here in order to provide an important overview.

A Broad Overview

By first examining the Earth's historical development, one obtains an appreciation for its complexity and the many dramatic changes it had undergone over the course of 4.5 billion years. Next, examine the slow evolutionary changes in the biological arena: Homosepiens appeared on earth only in very recent times. All progress of course, hinged on a supply of energy: internal energy (fission) from the Earth's core and a constant supply of external energy from the Sun. However, in order to effectively utilize the Sun's energy, a large quantity of converters was necessary. These converters were the photo synthesizers chlorophyll (a porphyrin), They must have been present, in one form or another on the Earth in its earliest history. The origin of porphyrin compounds, therefore, presents another interesting challenge. In fact, the origin of **all** organic molecules requires an extended analysis and evaluation. Based on recent studies, the Earth's organic molecules originated from two different sources: internally via hydrocarbons from the Earth's core, and externally from comets. Note the redundancy factor. Meanwhile, one is automatically drawn into the question of solar system development. Chapters 3 & 4 in this text described this complex and quite involved development: Our Sun gave birth to its planetary offspring. The Earth being one of the planets, went through its own development phases. The large gas planets produced their own off springs (satellites). The satellites of the major planets also went through their own development phases by producing comets. The comets in turn, produced the large assortment of organic molecules, occasionally dropping their content on our planet. Clearly a necessary, yet complex, multiple series, evolutionary development.

What did all of these complex elements have in common?

They all contained nuclear fission cores of different masses.

About 250–500 million years after the "birth" of the Earth, its size was approximately ½ its present radius. This can be shown readily by placing all the continents next to each other on a globe half its present size.

If all the present land areas coalesced to form a sphere of radius R_0 (reverse continental drift and shrink the Earth), then the value of R_0 can be calculated as follows:

$$\left(\frac{\textit{Land Area}}{\textit{Total Surface Area}}\right)^{\frac{1}{2}} = \left(\frac{1.48}{5.11}\right)^{\frac{1}{2}} = \frac{1}{1.86}$$

Hence $R_0 = 3430$ km, since $(1.86)3430 = $ (present radius) $= 6370$ km.

Note that the measured radius of the earth's high-pressure gas core is <u>3470 km</u>. (The constancy of this radius over a 4 billion time period is remarkable.)

Continental drift and expanding Earth models were challenged since they required high initial densities of 25–40 gm/cc and enormously large internal energy sources. A few brave scientists had already considered the presence of a fission reactor in the earth's core and its potential energy generation mechanism, such theories were rejected 7 years ago, to satisfy erroneous cosmology. (Degueldre & Fiorina 2016)

Life and Energy

All living and reproducing entities on Earth (which includes plants, trees, fish, reptiles, mammals, plankton, etc.) require energy in order to function. With the minor exception of geothermal heat energy, the ultimate origin of all this necessary energy is our Sun. The Sun's incoming energy (of about 100 Watts per square foot) is unable by

Table 6.1
Major Events in the Earth's History

Millions of Years Ago	Event
4600-4500	Origin of Earth and Moon's spin-off
~3800	Amitsoq & Kaapvaal gneisses, evidence that life is well established (carbon isotopic ratios)
~3400	First appearance of stromatolites (bacterial algal mats) in Onverwacht Group and Australia
1000-570	First appearance of ediacaran metazoans (soft body multicell animals), Skilogalee microbiota, Grenvilian Orogeny
430-400	First appearance of lungfish, land plants, jawed fish, North China rifts from Gondwana
400-380	First appearance of wingless insects, ferns, lapetus Ocean finally closed
320-300	First appearance of winged insects and pelycosaurs (early mammal-like reptile)
215-205	Early mammals, terminal Triassic extinction, Rifting between Gondwana and Laurasia
155-145	First appearance of birds, Paleo Tethys closed
95-85	Cretaceous anoxic event, Labrador Sea opens, India-Madagascar separate, Australia parts Antarctica
65-60	North Atlantic lavas, Indian Ocean spreads northwest of Seychelles, Yucatan Basin opens as Cuba moves north, Laramide Orogeny
60-55	First appearance of horses, grasses, and primates
25-20	Okinawa trough opens, Japanese Sea opens, Corsica-Sardinia parts France, East African and Red Sea rifting begins, Balearics/Khabalirs rift from Iberia
15-10	First appearance of hominids
3-0	Homo sapiens, Neanderthal man, Homo erectus, Australopithecus africanus, worldwide glaciations

itself to perform the required life functions without some intermediate conversion scheme. Hence, the availability of energy **converters** are absolutely necessary over the whole Earth's surface. Energy conversion is made possible by a very unique class of organic molecules known as porphyrins. Without porphyrins, life as we know it would not exist. The most primitive living organisms, the planktons, contain porphyrins. Planktons are the beginnings of a long food chain in the ocean, where one species eats up another species and so on. In general, the apex predators represent the largest species in the long food chain.

The Sun's energy at a specific wavelength gets efficiently absorbed by porphyrins, which in turn provide free electrons to enable the formation and development of cellular function and structure. These processes have been studied by chemists and biochemists and are amply documented in the scientific literature and textbooks. On Earth, these porphyrins are known as Chlorophyll A, B and C. The most primitive chlorophyll, C, is found in sea kelp.

Humans do not possess an outer skin containing chlorophyll. We therefore must ingest plants as food in order to receive our energy supply. By ingesting parts of animals that already have processed plants as food, (cows, for example) one can also receive energy. However, this extended process decreases the overall net energy utilization efficiency by a factor of 10. Thus, if we consider a group of people isolated within a finite area of arable and fertile land, the largest possible population size they could reach by consuming food

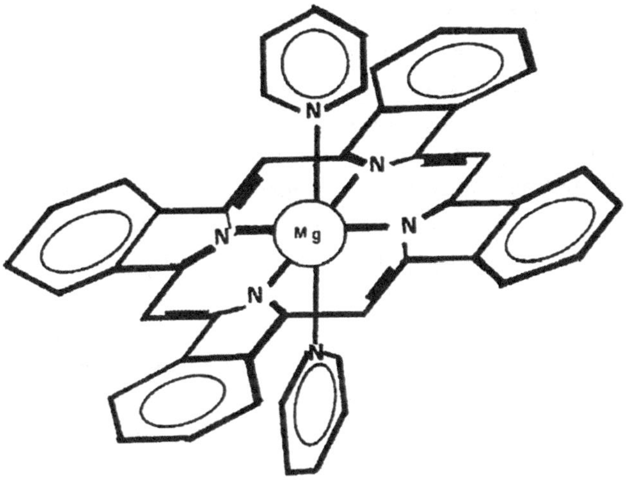

Figure 6.1

raised on their land would depend on whether they are vegetarians or meat-eaters. If their diet choice is vegetarian, this group has the potential to feed ***ten times*** as many people as if they ate only meat.

Since this text is intended to cover a broad perspective, we shall leave further discussions of this topic for readers to pursue more fully at their own leisure.

Now the question remains, "Where do these essential porphyrins come from and how are they produced?"

There is indirect scientific evidence that these porphyrins may also be present in other solar systems. A specific porphyrin, Magnesium tetrabenzoporphyrin (MgTBP), (discussed in Chapter 8) has been identified in the vast interstellar medium. The overall energy absorbing efficiency of this interstellar porphyrin is enormous. It also fluoresces with over 90% efficiency in the red spectral region and emits specific infrared frequency (known as UIR), which can be detected on Earth, using large infrared equipped telescopes. MgTBP also emits signals in the visible region. These signals are known as Diffuse Interstellar Bands (DIBs).

Low-temperature laboratory experiments have been able to mimic these DIBs signals and thus unambiguously identify MgTBP as the dominant molecule in the interstellar medium. (See chapt.8 for details). Examine the molecular structure of MgTBP (Figure 6.1). It resembles a miniature antenna. With eighteen conjugated bonds, free electrons can move easily around the outer structure of this molecule, which accounts for MgTBP's extremely high energy efficiency.

The Interstellar Medium

If you look up into the sky in the right direction on a clear night, away from city lights and smog, you will see the Milky Way. The Milky Way comprises billions of stars. From our distant vantage point inside the galaxy, we see some of our galaxy edge-on. The region where the sky appears brightest is a view along the edge of our galaxy. If you consider each star a raisin, we are indeed inside this raisin pudding looking out. Another intriguing part of our story concerns the makeup of the dust constituents within the clouds of the interstellar medium.

One of the enigmas associated with dust clouds are the chemical and physical details of its constituents. The earliest clues came from a study of the light scattering (reddening) process itself:

The beautiful, reddish glow of a sunset is created when small dust particles in the atmosphere scatter blue light, so that the remaining un-scattered light that reaches us is more reddish in appearance. Some of the more spectacular reddest-looking sunsets are seen when particulate matter (smoke from brush fires, or smog, for example) is very abundant in the atmosphere. The technical name for this scattering process is **Rayleigh Scattering**[1]. The shorter the light wavelength, the more readily is that light scattered. This is why our sky looks blue (the moon's sky is black because there is no atmosphere on the moon). Astronomers assume that similar processes occur in the interstellar medium

Gas atoms or small molecules do not contribute to the interstellar reddening phenomenon. The scattering process depends critically on the sizes of the particles, their chemical makeup (whether they are dielectric like ice or metallic), how much they absorb, and their shape. With such a host of variables, it is hopeless to find a molecule that explains the observed interstellar reddening unambiguously.

It is much easier to identify atoms and molecules in the stellar atmosphere! Whenever one studies atoms and molecules in stars, in the intervening interstellar medium, or in a laboratory, their characteristic signatures of emission and absorption spectra are clear, unambiguous, uncluttered, and easily identified. Such studies are now almost routine. Likewise, it has been possible to identify atoms and small radicals in the interstellar medium using similar absorption studies as was used for atoms and molecules. A list of optically identified atoms, radicals, and molecules is shown in Table 6.2. The sodium D, and calcium lines show up very clearly and can be used as reference markers.

Until 1940, only a limited set of interstellar diatomic molecules (radicals) and atoms were known. The situation remained unchanged until 1951, with the discovery of radio signals at a wavelength of 21 cm, emitted by excited hydrogen atoms throughout our galaxy. Since then, our entire galaxy has been carefully mapped using the 21 cm radio signals as markers. Hydrogen radio signals clearly outlined the spiral arms of our galaxy.

Table 6.2
Interstellar Atoms and Molecules Detected Optically

H, He, Ca, Na, K, Ti, Fe	Atoms
CH, CH, CN, CO, H_2	Radicals or molecules
Mg, $C_{46}H_{30}N_6$ (Molecule χ)	Largest interstellar molecule identified to date

The Interstellar Porphyrin Discovery

The author of this text had spent many years, on experimental spectroscopic efforts, searching for a molecule which would fit the interstellar spectral signatures or "fingerprints", known as the diffuse interstellar bands (DIBs). By the early 1950's, astronomers had discovered a set of these DIBs that did not correspond to any known atom or simple molecule. It proved to be a formidable challenge, since, ordinarily, one has ample laboratory spectra immediately at hand to compare with the spectra taken by astronomical means. The clues leading to its molecular nature were the width of its spectral lines. The designation "diffuse" meant that the lines were wide (some as wide as 20 Å), whereas atomic lines are ordinarily about 1/10 Å wide. Wide lines are usually associated with solids, liquids, or large molecules. The problem was baffling since some of the diffuse lines were also relatively narrow. How does one get both wide as well as narrow lines? These were the horns of the dilemma.

Since other astrophysicists, engaged in similar searches, had no doubt exhausted the study of simple molecules, particularly molecules extending up to 12 atoms, this author decided to make a drastic jump into the realm of still larger molecules, whose spectra perhaps had not been measured or catalogued or were taken under very specific conditions. All three assumptions proved to apply to the molecule ultimately identified as responsible for the DIBs. The dilemma was still, however, a "needle in-the-haystack search": to seek the correct molecule out of a total of hundreds of millions of possible compounds. All clues were systematically explored, a great deal of hard work had to be done, and a certain amount of luck was needed.

The author and two lab assistants examined spectroscopically thousands of compounds. A search was made of every available spectral catalogue. The most valuable clue proved to be the fact that one of the diffuse interstellar absorption lines at 4,428 Å is the dominant one. Also, certain regularities in the spectrum were suggestive of vibronic spacings associated with well-known molecules.

In 1967, it became apparent that porphyrins might be potential candidates, and in 1969, one particular porphyrin (Figure 6.1), magnesium-tetrabenzoporphyrin proved to be the correct one. It has its strongest spectral line exactly at 4,428 Å, it has no other lines stronger than 4,428 Å, and its line width was comparable of the astronomical data. Its second strongest line, at 6,284 Å, was also comparable in intensity. Four additional lines, were also found, making a total of six coincidences as of 1975. (Spectroscopic Studies of Interstellar Grains, NASA CR 1677, 1970.) Many years later, with university student assistance and over a hundred experiments, 90 DIBs were spectroscopically matched with lab data. (Johnson 2006) See Chapt.8 for details.

The presence of porphyrin molecules in outer space is clearly significant. Porphyrins have been found in carbonaceous chondrite meteorites by Dr. Gordon Hodgson in 1969. Porphyrins are also found in petroleum (to the chagrin of the oil industry) and in very ancient shale sediments, which are at least a few billion years old.

Porphyrins have many unique properties. They have thermodynamic stability, which allows them to survive billions of years. They are important for prebiotic evolutionary theories, since they have the ability to absorb sunlight and produce free electrons. Protoporphyrin IX probably constitutes one of the earliest and most significant porphyrins, since it converts sunlight into another form of energy.

The Microwave Minimolecules

Since 1969, approximately 200 "mini-organic" molecules have been discovered in the interstellar clouds using microwave techniques. Some of these molecules have as many as eleven atoms (to be contrasted with the 83 atoms in the interstellar porphyrin molecule (MgTBP)). The discovery of microwave mini molecules proved fortunate, since it made it easier to accept the presence of larger organic molecules in outer space. The physical properties most often questioned about the porphyrin molecule, are its stability and its mode of production. In terms of stability, it turns out that a porphyrin molecule is more thermodynamically stable than 90% of all other "mini-organic" molecules. The lab experimental breakup of MgTBP by means of electron-bombardment resulted in the production of smaller molecular fragments. These disintegration products of MgTBP contain combinations of the atoms C, H, and N, and have in fact been discovered in interstellar clouds with microwave detection techniques (19 so far). Moreover, the laboratory ratios of these disintegration products are in good agreement with the observed astronomical abundances of these molecules (Table 6.3).

Table 6.3
The Disintegration Products of MgTBP by 70 e.v. Electron Bombardment—a Few Selected for Astrophysical Interest

Molecule	Abundance (Lab.) (percentage)*	Astronomical Abundance**
HC_3N	11.1	2×10^{16} cm^{-2}
CH_3CN	0.1	2×10^{14} cm^{-2}
CH_3C_2H	0.24	

*F.M. Johnson, D. Rosenthal, Bull. A,. Astron. Soc. 5 (4) (1973) 422.

**(Townes R. & Welch 1971)

This author's 2006 publication (Spectrochimica Acta Part A **65**, 2006, 1154–1179) is to date, the most complete, published exposition of the interstellar porphyrin solution, plus other important ramifications.

Discussed more fully in Chapter 8 is the reason why it had been so difficult to completely unravel the totality of the DIB astronomical spectra: Mother Nature was playing a cruel trick. The observed DIB spectral bands appear mainly in *absorption* (some rare DIBs are seen as fluorescence) since they originate via a rather sophisticated fluorescence multiple scattering process in fluffy cocoons composed of paraffin/pyridine mixtures. Lab data does not readily reproduce this unexpected process. MgTBP is a proven, highly efficient fluorescent emitter.

Another important clue was the spectroscopic lab identification of an interstellar UV absorption band at 2,175 Å. It turned out to be the critical solvent molecule pyridine, whose *transmission* window is exactly at that wavelength. Light is scattered within a fluffy paraffin cocoon, such that the background stellar radiation produces a *missing* wavelength, making it look like absorption (whereas it is in fact a transmission window). In other words, it resembles a fictitious absorption band and thereby continues to mislead most astronomers, (another one of Mother Nature's tricks).

Mysterious Infra-red Interstellar Emission Signals—Identified

The earth's atmosphere acts a selective screen preventing us from seeing ultraviolet and a swath of incoming infrared (IR) signals. However, as soon as advanced technology allowed us to penetrate this atmospheric barrier (via balloons, rockets and satellites) we were able to detect and measure those previously inaccessible electromagnetic signals. For example, NASA equipped a 747 airplane with a telescope and IR detection equipment. By flying at extremely high altitudes, this plane allowed scientists onboard to collect a slew of diverse IR emission signals. These signals were cataloged as "unidentified infrared bands" (UIR). Their identity was established and described in detail in chapter 8.

References

Degueldre, C. & Fiorina, C. 2016, Solid Earth Sciences, Volume 1, 49–63.

Johnson, F.M. 1972, *Mem. Soc. Roy. des Sciences de Liege*, 611 series, Vol. III, 39 -407.

Johnson, F.M. 2006, *Spectrochimica Acta A*, 65, 1154–1179.

Townes, R. & Welch 1971, *Science*, 774, 1083.

Chapter 7

Gravity - Its Real Origin

Abstract

Gravity is the aggregate of leaked nuclear forces, arising from quantum-mechanical tunnelling emanating from the cores of atomic nuclei.

Introduction

According to Sir Isaac Newton, (1642-1727) the concept of gravity involves **forces.** In 1665, Newton, at the age of 23, discovered the law of gravity which he defined as follows: Every object attracts every other object with a force that is in direct proportion to the product of the masses of each object and inversely proportional to the square of their distance separation.{1} The strength of this force is defined by a constant of proportionality **G**, called the gravitational constant. (The inverse square distance law merely implies that forces between masses extend symmetrically in all directions). The earliest experimental measurement of G, of this very weak gravitational force, was performed by Henry Cavendish in 1797-1798 at Cambridge University. Despite the importance of G in such fields as geophysics, theoretical physics, and astrophysics, the measurement precision of G has only improved by two orders of magnitude in 200 years. In fact, despite a large number modern, highly complex G experiments to date, it is still the least precise measurement, among all of our other fundamental physical constants. Moreover, a list of 13 different precision G measurements range from 6.672 to $6.676 \times 10^{-11}\ m^3kg^{-1}gs^{-2}$. The apparent large range in G values is inconsistent with respect to the high precision of each individual G measurement [2]. This suggests that some basic physical aspects are being overlooked. Indeed, Important, almost overlooked experiments by Eötvös [3] provide the clue to this enigma.

Einstein's General Relativity, (GR), discovered in 1915, was far more complex. It implied that gravity arises from the **warping of spacetime**. Einstein developed a tensor gravitational field equation, which consists of ten different equations and incorporates an energy-momentum, or energy-matter tensor, relying on the Principle of Equivalence and curvature of space-time [4].

GR overlooked any details of matter: It was a fundamentally different approach to gravity's origin. Moreover, Eötvös' experiments, published a century ago [3] had shown minor deviations from the weak equivalence principle (WEP), which were either ignored or generally overlooked by others. Eötvös had meticulously measured the net torque, from the arms of a torsion balance, between a set of samples which could have different inertial (m_I) and gravitational mass (m_g). If one defines deviations from WEP as $k = (m_g)/(m_I) - 1$, then Eötvös' measured deviations(Δk) ranged from $4 +/- 1$ to $-10 +/- 2 \times (10^{-9})$.

Eötvös' eleven experiments had shown 7 pairs of substances deviating from the Weak Equivalence Principle: one of them, by as much as **5σ** from an expected null result. According to Fischbach [5], the probability that this difference could have arisen by chance was 1 in 3.5 million.

In view of our most recent discovery regarding our solar system's evolution,[6], the role of neutrons and their initial entanglement has to be considered. Neutrons in turn can mutate into protons, electrons and neutrinos, where each one of them has a spin of $h/2\pi$. (All matter is made up of atoms, which in turn are composed of protons, neutrons and electrons).

Within each nucleus, of all known atoms, neutrons and protons, are held together by enormously large forces (about 8 Mev per nuclear pair). These pp, nn and pn forces are confined to an extremely small radius of about 1.5×10^{-13} cm. Scientists assume that these forces are totally confined within the nucleus of each atom. The possibility that a minute fraction of these forces could undergo Quantum- Mechanical Tunnelling, has never been under consideration.

Modern Gravity

By means of quantum mechanical tunnelling, a minute fraction of nuclear forces escapes from all nuclei. THIS BECOMES THE UNIVERSAL ORIGIN of GRAVITY.

The "tunneled" nuclear emission radiates symmetrically in all directions (the inverse square relationship being satisfied). Meanwhile, the escaped "attractive" force- energy component seeks to combine with any distant proton or neutron, at a very low probability. Hence the G factor, which is a measure of this illusive force is extremely small.

Since it is not an electric field, the small force which escapes a nuclear core does neither affect molecules nor electrons. **Gravity is a nuclear force**: It is not a separate force nor an elusive fifth force.

Consistency related to Newton's law of gravity.

1. The leaked nuclear force from each atom is emitted symmetrically in all possible directions-hence it obeys an inverse square distance relationship.
2. The total tunneled (leaked) nuclear forces from each atom are proportional to the mass of their originating nuclei. These forces must be absorbed in other atoms to produce an attractive force between them.

Evidence for the Leaked Nuclear Forces

In 1986 Fischbach et al. [7] reanalyzed Eötvös' classic 1922 paper, which measured and compared the gravitational and inertial content of eleven different substances. They introduced a parameter (P) that was assigned to each of the substances, in Eötvös' experiment. The parameter (P) is the baryonic charge -to-mass ratio (B/μ), where B is the total baryon number of each sample and $\mu = m/m_H$, where m is the sample mass and m_H is the mass of hydrogen = 1.007825u, u being the atomic mass unit. The baryon number is the sum of protons and neutrons in the nucleus. The parameter (P) is a measure of the barrier height of a nucleus. Quantum mechanical barrier leakage depends on barrier height. Comparing a pair of different substances, the one with the lowest barrier would have the greater leakage. Violations of the weak Equivalence Principle (WEP) were defined by $k=(m_g)/(m_I) - 1$. Eötvös' experiment measured differences in k for a variety of combinations of substances. Fischbach and Talmadge [8] presented a Table which listed each of the eleven sample combinations against their Δk and $\Delta(B/\mu)$ values, (Table1) They then plotted each of their entries in a graph, (Fig.1), which displayed a best fit, **linear relationship** among the eleven experimental entries. Since the leaked nuclear forces are barrier height dependent, this becomes the proof that the **correct interpretation for Eötvös' experiments is quantum mechanical tunneling of a small amount of nuclear forces from the cores of atomic nuclei.** (It also confirmed a violation of the WEP).

Discussion

I am grateful that Fischbach and Associates were persistent in their interest in Eotvos' experiments. Although Fischbach's interest was in search for finding a fifth force, he unknowingly found the evidence from these century old experiments for Gravity being part of the **third** force.

Samples Compared	$10^9\Delta\kappa$	$10^4\Delta(B/\mu)$
Magnalium-Pt	4 ± 1	+5.00
Brass-Pt	1 ± 2	+9.32
Cu-Pt	4 ± 2	+9.42
Ag-Fe-O_4-Pt	0 ± 2	0.0
$CuSO_4$ (dissolution)-Pt	2 ± 2	0.0
Snakewood-Pt	-1 ± 2	−5.09
Asbestos-Cu	-3 ± 2	−7.40
$CuSO_4 \cdot 5H_2O$-Cu	-5 ± 2	−8.57
$CuSO_4$ (solution)-Cu	-7 ± 2	−14.63
H_2O-Cu	-10 ± 2	−17.18
Tallow-Cu	-6 ± 2	−20.31

Table 1 (from ref.8) Values of Δk and Δ(B/μ) for the various combinations of samples used in the Eötvös' experiment.

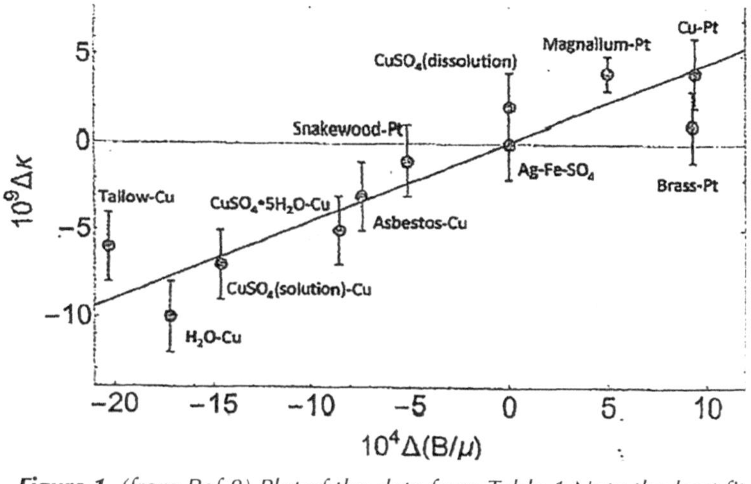

Figure 1 (from Ref.8) Plot of the data from Table 1 Note the best fit linear relationship of Δk vs Δ(B/μ).

Gravity is most likely not universal but is unique for each solar system. This is because all matter in our solar system originated from the spinning, Sun's core, entangled neutrons. 4.5 billion years ago, these neutrons were involved in the production of our solar system, which sequentially spun material outwards to produce our planetary system [6].

The same method also produced SNFRs (spinning nuclear fission reactors) from a rapidly spinning, solar neutron core. This favors the production of atoms having the largest number of neutrons. This in turn produces atoms of both stable and unstable variety, which would include Uranium 238 and 235 atoms plus fast and slow neutrons. Thus, fission reactors are produced on the Sun's spinning core. The abundance of hydrogen is the result of Beta decay, while helium is the result of Alpha emission from atoms with large A and Z values.

Conclusion

The four forces in the universe that theorists consider are: Electromagnetism, strong nuclear forces, weak nuclear forces and gravity. These can now be simplified to: Strong, weak and very weak nuclear forces plus electromagnetism; or simply, to only two forces: nuclear forces and electromagnetism.

SPINS and Gravity Repulsion

There are three possible types of exchange forces operating within a nucleus: [9]

Heisenberg forces: in which there is exchange of both the position and spin coordinates of two interacting nucleons. They are attractive for triplet interactions and repulsive for singlet interactions (antiparallel spins).

Majorana forces: there is exchange for position coordinates but not for spin.

Bartlett forces: Exchange for spin but not for position coordinate. A huge mixture of unsorted atomic nuclei would generate forces which favor attractive forces. If one wishes to tailor specific force requirements, careful selection of atomic isotopes is necessary.

There are also differences in the types of forces for each of nn, pp, and pn pairs based on quantum mechanics. Nuclear particle spins also play an essential role in view of their quantum mechanical importance. Gravity repulsion therefore requires both careful isotope selection, plus particle alignment on an expanded scale. Molecular beam and magnetic field technology are necessary in order to sort out the required isotopes for each specific application.

A huge mixture of unsorted atomic nuclei would generate forces which favor attractive forces. If one wishes to tailor specific force requirements, careful selection of atomic isotopes is necessary.

This in turn would blend in with the quantum mechanical exchange force selection which produced a 5000 times amplification in the ferromagnetic analogue case.

References

[1] I. Newton,Philosophiae Naturalis Principia Mathematica, 1687

[2] Chao Xue et al. *Precision Measurement of the Newtonian Gravitational constant,* National Science Review, **7**, Issue 12, 1803-1817,D ec.2020.

[3] Eötvös, R, Pekar,D and E. Fekete, Beiträge zum Gesetze der Proportionalität von Trägheit und Gravität, Ann. Phys. (Leipzig) **68** (1922) 11.

Z. Szabo, *Three Fundamental Papers of Lorand Eötvös,* Lorand Eötvös Geophysical Institute of Hungary, Budapest,1998

[4] A. Einstein, *The Field Equations of General Relativity,* Nov25,1915, Sitzung Berichte of the Prussian Academy of Sciences

[5] E. Fischbach and D.E.Krause, *The Eötvös Paradox: The Enduring Significance of Eötvös' most Famous Experiment,* Proceedings of Science, **353,** Febr18, 2020

[6] F.M. Johnson, *Recent Discoveries in Astronomy, Forensic Astronomy* (2021), Lawyers and Judges Publishing Co,,Inc.

F.M.Johnson, *Planetary Cores, their Energy Flux Relationship, and its Implications,* Planetary and Space Science **151,** 125-129, 2018.

[7] E. Fischbach et al. *Reanalysis of the Eötvös Experiment,* Phys.Rev.Lett. **56,** 3, 1986

[8] E. Fischbach and C.L. Talmadge, "*The Search for Non-Newtonian Gravity*' AIP/Springer Publi.Co., New York,1999

[9] R.D. Evans, *The Atomic Nucleus, McGraw* -Hill Book Co., Inc. 1955

Chapter 8

The Solution to a Unique Spectroscopic Enigma

A seemingly large, intractable, spectroscopic problem, the diffuse interstellar bands (DIBs) is found to have a unique solution, once preconceived ideas are replaced by a set of unexpected yet essential conditions. A reexamination of (mistaken) assumptions regarding specific telescopic (DIB) observations was one of the many necessary steps which played an important role in resolving this enigma. Furthermore, towards Star HD 183143 sightline a total of 440 DIBs (Hobbs *et al.* 2009) are seen. Surprisingly, there are 375 DIB pairs among this set of DIBs, having seven different splittings. The significance and implications of this discovery contribute in part towards the solution.

I. Introduction

This chapter provides an overview of a long-standing spectroscopic challenge (with its preconceived ideas) and concludes with a detailed step-wise analysis towards a comprehensive solution. The enigma relates to over 500 individual spectral bands (Fan *et al.* 2017) (seen towards specific stellar backgrounds), whose diverse line widths exceed those usually seen in atomic and molecular spectra. Heger (1922) was the first astronomer to discover two such examples of broad lines. Merrill (1936) discovered four more broad lines and named them diffuse interstellar bands (DIBs). Their origin was established to be interstellar (Merrill 1934, 1936). The historical aspects concerning DIBs are included in a list of over 100 references by Snow (2013). His article also provides the dominant perspective on how the DIB enigma is to be perceived, and the methodology by which it is presently pursued. There were two conferences devoted exclusively to DIBs: the first in Boulder, Colorado (1995, Tielens and Snow Eds.), and the second one was sponsored by the IAU (2013). Despite many misgivings, and a plethora of distractions, a unique DIB solution does indeed exist. However, it requires a myriad of essential conditions which are enumerate here. Contrary to expectations, it is a unique molecule that eventually solves the enigma. It was already identified in 1969 (see Johnson 1972). The condition and the specific, unusual circumstances by which such a molecule could produce over 500 DIBs requires an extended analysis and discussion, plus a set of special conditions, which were not generally recognized until recently. The reader should be prepared for many surprises regarding an adventure into the content of the mysterious interstellar medium (ISM).

II. Spectral Measurements

A. Background

Absorption and emission lines from solar as well as stellar sources, using telescopic/spectroscopic combination technology have been studied for over a century. During the first few decades of the 20th century, optical astronomical observations included a study of the Interstellar medium (ISM) where the discovery of various atomic elements such as hydrogen, helium, calcium, sodium, potassium and iron were augmented by the identification of diatomic species such as CH, CH^+, CN and OH (post 1969, using microwave technology exclusively). This list was substantially enlarged to include more than 170 diverse molecules and molecular fragments (Irvine *et al.* 1991 and Cox 1999). Whereas atomic and molecular spectral lines have widths of about $1cm^{-1}$, Heger (1922) inadvertently discovered two ISM spectral features at 5780 Å and 5797 Å whose line widths were much wider (several cm^{-1}). P. Merrill (1936) discovered four more such features (Merrill and Wilson,1938). Herbig (1995) extended this number to over 100. The most recent DIB Atlases by Hobbs *et al.* (2008, 2009) and Fan *et al.* (2017) list over 500 such enigmatic DIBs. The DIB enigma is alleged to be "the longest-standing unsolved problem in the history of spectroscopy" (Snow 2013). This chapter will address and negate this remark while simultaneously providing a multi- stepwise complete solution to this enigma.

B. Measurement Methodology

Each of the many facets related to this complex problem would however have to be explored in detail: Distant starlight viewed through translucent clouds in the ISM becomes reddened due to Mie scattering, measured by a term called extinction [$E_{(V-B)}$]. V and B are optical intensity measurements in specific visible (V) and blue (B) regions of the spectrum. DIB intensity measurements (with some exceptions) are generally found to be related to its corresponding $E_{(V-B)}$ values, towards reddened starlight in a specific line of sight. Moreover, a change in the slope of the extinction curve (near 4400 Å was noted by (Whitford (1958), Nandy (1964), Underhill and Walker (1966). It was experimentally confirmed by (F.M. Johnson 1970, Pg. 73). This would constitute **Clue #1**, which implies the existence of a direct relationship between the DIB carrier and its immediate environment.

An extensive search of the spectroscopic literature, coupled with an equally elaborate and wide-ranging experimental program, resulted in the discovery of porphyrins as the most likely DIB candidates (Johnson 1970). Charles Castro synthesized a set of three tetrabenzoporphyrins: MgTBP, FeTBP and H_2TBP. This was followed by a thorough exploratory program of these molecules at room and liquid nitrogen temperatures using a diverse set of solvents (Johnson 1970, and Ehrenberg & Johnson 1992). **30** experimental absorption and fluorescence spectral features confirmed the presence of MgTBP (Johnson 1972, Tables 1 & 2). Each spectral band either matches a known or a (hitherto-fore unknown in 1971), new DIB in Atlases by (Hobbs *et al.* 2009 and Fan *et al.* 2017). The veracity of these results was unfairly criticized by Donn & Khanna (1980). Inept experimental procedures were employed by Donn at room and at 77K temperatures: An unorthodox substance, a nujol mull, was used in the optical spectral region for absorption measurements of MgTBP. Seriously flawed experimentation could obviously not reproduce any of Johnson's (1972) spectral data.

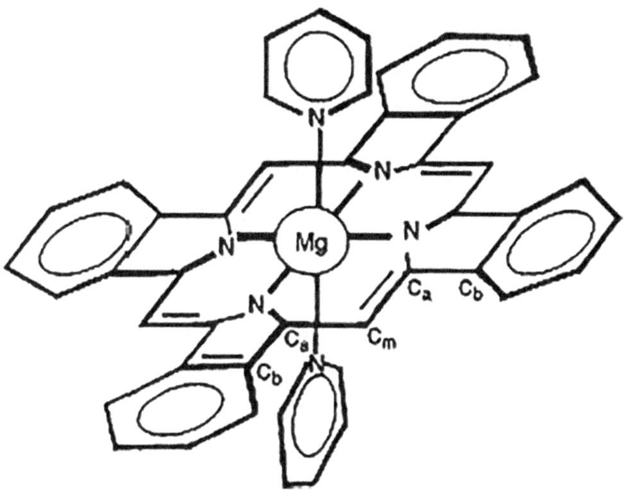

Figure 1 *Dipyridyl magnesium tetrabenzoporphyrin* ($MgC_{46}H_{30}N_6$)

III. The Key DIB Molecule: Magnesiumtetrabenzoporphyrin (MgTBP) (Clue #2)

MgTBP produces a brilliant red florescence on excitation by blue light, at a conversion efficiency of about 90%. It sublimes at 500 C and is probably one of the most thermodynamically stable, organic molecules in nature. Furthermore, its molecular extinction coefficient of its dominant 4428 Å absorption feature (Soret band) is one and a half million, [f ≥ 1], which enables a determination of its interstellar abundance]. Low temperature experiments used a high intensity tungsten lamp, a HeNe, argon ion and/or HeCd (4416 Å) laser. Spectroscopic data was collected by means of a 3/4 m Jarrell Ash spectrometer (dispersion of 20 Å/mm) and was recorded on 5" x 7" photographic plates. For further details, see Johnson 1970, 1995 and 2006.

IV. The Shpolskii Matrix (Clue #3)

With the exception of the Red Rectangle (Van Winckel *et al.* 2002, 2013), the DIB-Carrier (MgTBP) is not in a gas phase nor embedded within a solid matrix.

In the ISM, it is situated in a unique intermediate state, surrounded by a very loose grouping of mainly paraffin molecules, e.g. octane or decane, which prevent its free molecular rotation. However, they do allow specific vibrational frequencies, depending on solvation and the detailed location of its enveloping paraffin structure, whose diverse variations are designated as "sites". This unique, effective porphyrin isolation technology was discovered by E.V. Shpolskii, (1952, 1962 A & B), Sevchenko (1965, 1967) and Zaleski (1973, 1975). It facilitates measurements of vibronic transitions in these molecules (see extended list in (Johnson 2006). Wm H Smith (1995) suggested the possibility that the DIB carrier "could be trapped within a cage like structure which would prevent dissociative processes". Smith's proposal is somewhat similar to the type of fluffy paraffin matrix envisioned as a Shpolskii matrix. The preferred solvent is pyridine, with two of them symmetrically attached to the central magnesium atom of MgTBP (see Fig.1). The presence of pyridine as solvent is experimentally confirmed by the UV hump at 2175 Å (Johnson2006).

V. DIB Observations (Clue #4)

DIB Telescope Observations

DIB telescopic observations are generally seen as diffuse ***absorption** bands* in the spectra from specific distant stellar background light sources, transmitted via multiple scattering of paraffins which surround the DIB carrier (MgTBP). Hence, of DIB's total **fluorescence,** only a very bright fluorescence emission band, such as the 6614 Å DIB, would have sufficient intensity to have some of its fluorescence survive the multiple scattering process and be observed as a DIB emission signal (Williams *et al.* 2015).

A. The Unique Red Rectangle

This highly unusual object consists of a binary central star HD 44179 surrounded by a so-called "circumbinary disc." Its diverse emissions include spectra of an Extended Red Emission (ERE) whose maximum intensity is in the 6370Å region (Van Winckel *et al.* 2002), blue spectral emission features extending from 3700 Å to about 4500 Å (Schmidt *et al.* 1980), superimposed substructures consisting of a complex succession of narrow **emission** features (Van Winckel *et al.* 2002) plus a set of **UIR** emission bands (Russell *et al.* 1978). The uniqueness of the Red Rectangle is its ability to display the DIB carrier molecule **MgTBP** *in all of its development stages*. This would include its gas phase, based on experimental data by Even *et al.* (1982 and Even 1989) analyzed by F.M. Johnson (2006). Its multiple red **emission** bands correspond to the observed **DIB absorption** features (Scarrott 1991, 1992). **The RR thus confirms that in the ISM, a substantial number of DIBs seen in absorption are in fact fluorescent signals,** trapped within paraffin cages of varying complexity. Moreover, the RR spectacularly demonstrates the successive stages of "cage" buildup, (simulating some observed DIB spectra). As the buildup of successively larger size molecular cages continues, it will ultimately produce an inhomogeneous medium for MgTBP and thereby result in the production of the ERE. This result can easily be duplicated with MgTBP in a laboratory setup at 77K, including a display of its approximate, maximum intensity at 6350Å.

Fluorescence radiation is trapped within "Shpolskii" cage matrixes. Its long wavelength radiation does, however, readily escape these cages and is observed astronomically as the well-known UIR. Note that in Table XI, all the long wavelength infrared fluorescence signals are accounted for by the single molecule MgTBP.

We are also most likely witnessing the final stages of solar system development (F.M. Johnson 2015) which would involve the ejection of comets from major planets. Comets, in turn, are involved in the chemistry production phase, via hot plasma jet ejection processes (Hodgson 1967, 1972, 1973). MgTBP would be among compounds which could survive the high internal comet temperatures in the ejection process; thusremain intact while exiting the comet with an initial high kinetic velocity component which would allow it to leave a solar system and eventually be part of the ISM.

Interstellar gas clouds are the left-over refuse from prior solar system developments. They do not constitute the beginnings of planetary developments (as is erroneously still believed in 2023).

B. Theory

The basic MgTBP chromophore has spatial symmetry which places it in the point group D_{4h}.(Gouterman 1961) predicts a four-orbital model having a degenerate $S_1 \leftarrow S_0$ (Q-band) and $S_2 \leftarrow S_0$ (Soret band). The degeneracy of the excited states in MgTBP can be lifted by the action of a crystalline field originating within a Shpolskii host matrix or in a single paraffin crystal (Platenkamp and Canters 1981). The strong Soret and Q bands arise in part from the delocalized nature of the Π electrons shared by the 16 or 18 conjugated bonds in the ring (Figure 1). The Q-band transition is made possible by configuration interaction mixing. It is this unique configuration of shared Π electrons which produces the strong optical absorption and emission bands. Splitting magnitudes of MgTBP's excited electronic states depend on its location relative to a specific site. In the ISM, MgTBP is found to exist in at least 13 different sites with 24 origins and 7 different induced electric field splittings of DIB pairs for sightline towards HD 183143.

C. The DIB Pair Splittings

So far, unnoticed, is the surprisingly large grand total of **375** DIB pairs towards sight-line 183143. This constitutes a total of 750 DIBs which participate in pairing, whereas (Hobbs *et al* 2009) lists a total of merely 414 DIBs. This implies the existence of considerable spectral line overlapping, resulting in either asymmetric DIB line shapes or multiple complexities (misinterpreted by some as potential rotational molecular structure of unknown specificity). Herbig (1988, 1995) and Bondar *et al.* (2013) had noticed the presence of about half dozen DIB pairs. However, no one else seemed to have noticed their existence, their underlying significance, nor suspected the immensity of the total number of DIB pairs (see Table 3). Following in Herzberg's (1955, 1967 and 1977) footsteps, DIB data is primarily searched for possible incriminating evidence of molecular rotational spectra; yet none was ever found in any of the more recent studies (Hobbs *et al.* 2008, 2009, Fan *et al.* 2017).

VI. Discoveries By W.W. Duley and Stanislav Kuzmin (D & K) (2010) and Its Implications (Clue #5)

Herbig (1988) as well as Krelowski *et al.* (1997) had discovered in the 6800 Å region a series of some spectral lines separated by about 35 cm^{-1}. Using the DIB database of Hobbs *et al.* (2009) **D&K** revealed the existence of an additional organized spectral structure taking the form of extensive vibronic progressions with frequencies $a = 5.18$ cm^{-1}, $b = 24.41$ cm^{-1}, $c = 31.55$ cm^{-1}, and $d = 34.02$ cm^{-1}, by assigning the origins at 6862.61 Å and 6843,6 Å . The major significance of their discoveries is the following:

1) DIB vibronic assignments in **D&K's** Table1 display 29 individual energy shifts (the difference in energy between an observed DIB and its origin at 6862.61 Å). Each one of the listed shifts corresponds to a known vibrational energy of MgTBP (Table 1a, Johnson2006).
2) The 68 DIBs in Table 1 (D&K) produce a tight-fitting progression which implies that it must originate from a **specific single molecule** (rather than from a multiplicity of different molecules).
3) In **D&K's** Fig1, vibronic progressions display 16 potential DIB pairs, with energy separations (splits) of 40 cm^{-1}, origins at 6862 Å and 6843 Å respectively. (See clue #4)
4) The most striking progression involved the 21.41 cm^{-1} mode with origin at 6862.61 Å.
5) D&K point out that" Low energy structural modes (such as 21.4 cm^{-1}) are common in large carbon-based systems and arise from the type of collective vibrations such as drum-head oscillations (Rossing & Fletcher 2004)", see also Clue #11—**UIR**. These drumhead vibrations (of 20 cm^{-1} energy) originate from the central magnesium atom of MgTBP, vibrating perpendicular to the plane of the molecule. Thus, the 20 cm^{-1} vibrational frequency is an important oscillation, which has also been independently confirmed by a direct microwave DIB absorption observation by Műller, *et al.* (2013)

MgTBP HAS MORE THAN 200 VIBRATIONAL FREQUENCIES (CLUE #6) listed in Table 1 (Johnson 2006)

MgTBP with 2 attached pyridine ligands is situated within a matrix occupying a large number of possible (different) sites (Table3). Coupled with its large pool of potential vibrational frequencies, it is capable to produce the large variety of observed absorption/fluorescence spectra. In order to locate experimentally any one of these elusive potential sites, a minimum of a hundred different lab experiments were necessary, resulting in the production of 90 DIB-Lab spectroscopic matches (see Table4). A summary of DIB pairs and their origins are listed in Table 3. Origins are identified by virtue of their relatively larger EW values. For example, the most readily identified DIB pairs (6284 Å and 6270 Å, EW = 1884, 256) are split by 35-36 cm^{-1} and result in a total of **76 DIB pairs** (see Table 5). The DIB splitting magnitude is a direct measure of the electric field strength acting on the DIB carrier (MgTBP).

VII. Schematic Energy Level Diagram (Clue #7)

This diagram illustrates absorption and fluorescence DIB spectra processes. The splitting of the electric S_1 and S_2 states with its superimposed vibrational manifold indicates absorption into its excited states and their subsequent fluorescence from their Q_x and Q_y origins.

Johnson (2006) Table 1 provides a complete list of vibrational frequencies of MgTBP in its vapor phase and at

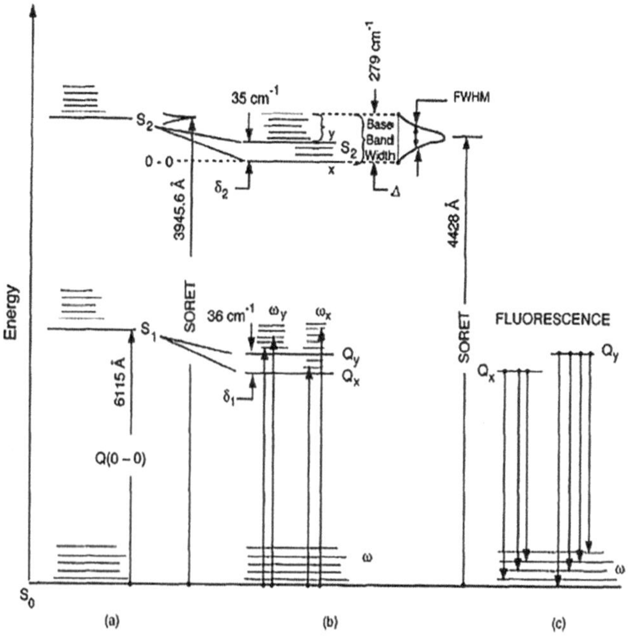

Figure 2 *Schematic energy level diagram, showing vibronic transitions in absorption and in emission for MgTBP.*

Note: the transition from the "bare" state to the matrix and the splitting of the S_1 *and the* S_2 *states.*

4°K in a paraffin matrix. For comparison, similar data is also shown in this table for molecules ZnTBP, CdTBP, H_2TBP and MgP. Table 1 also includes FTIR and RAMAN data.

Tables 5 to 10 lists virtually all of the DIBs (Hobbs *et al.* 2009) their respective diverse origins and the resulting energy differences between origins and each of the DIB's energy (cm^{-1}). These differences are confronted with the known vibration energies of MgTBP (Table 1, Johnson 2006). Note the overwhelming confirmation achieved by assigning each individual DIB to a specific, documented, vibrational frequency of MgTBP.

VIII. The UIR Spectra (Clue #8)

DIBs prolific optical fluorescence is largely trapped within its surrounding paraffin matrix. However, the resulting long-wavelength infra-red energy does escape and is observed in emission as the so called Unidentified Infra-red Radiation (UIR). Table 11 displays a list of thirty observed UIR emission bands (Peeters 2013) and its corresponding MgTBP vibrational frequencies. With the exception of 3 uncertain entries, note the almost complete identification of each of the 26 Infra-red emission signals, based on known MgTBP's vibrational frequencies.

IX. The Red and NIR DIBs (Clue #9)

The Triplet state origin (Table IX) at 8620.99 Å (11600 cm^{-1}) produces a set of 5 fluorescence DIBs (observed in absorption). The energy difference of its origin minus each DIB's energy[1] corresponds to a vibrational frequency[1] of MgTBP (Table 1, Johnson 2006). Note that 2 of the 5 fluorescent DIBs (**9632 Å and 9577 Å**) were erroneously assigned to 2 non-fluorescent fullerene (C_{60}^+ ion) absorption lines (Galazutdinov *et al.* 2016). In fact, other postulated DIB candidates (such as PAHs for example) can also be eliminated, merely on the basis of being non-fluorescent. The triplet state in ZnTBP (Fielding and Mau 1976) is similar to the MgTBP molecule.

All 30 DIBs (Table10) extending from 10360 Å to 17930 Å are (without a single exception) explicable by demonstrating the various vibrational frequencies of MgTBP listed in Table 1 (Johnson 2006), which include the longest wavelength DIBs discovered by (T. Geballe 2011, 2013) with its origin at 15273 Å. MgTBP's vibrational frequency at 238 cm^{-1} is involved in the 15850 Å DIB (observed by T. Geballe). Johnson's Table 1 gives 9 different, independent experimental methods for establishing this unique 238 cm^{-1} frequency.

[1] Energy and frequency, despite their ambiguous terminology, are always in units of cm^{-1}.

X. The Blue DIBs (Clue #10)

a) The origins of the split S_2 energy levels are established by lab data of Table 3 (Johnson 2006). They are 22444 cm^{-1} and 22479 cm^{-1} respectively. Sixteen of the Blue DIBs from (Table 12 Fan *et al*, 2017) converted to energy states (cm^{-1}) are each confronted by energy differences between each origin and its corresponding energy state entry. Note that each of these energy differences corresponds to a vibration frequency in (Table 1, Johnson 2006).

b) DIBs from Comet Hyakutake and Comet 122 P/de Vico are analyzed using the same origins as in (a) above. Note that the resulting energy differences are consistent with their assignment to vibrational frequencies of MgTBP (Johnson 2006 Table 1). A number of DIBs from these solar system comets are new and not listed in the standard Atlas (Fan *et al*, 2017). However, they are consistent with known vibrational frequencies listed in (Table 1, Johnson 2006). Vibrational frequencies common to both comets are identified by an asterisk*.

2) **Note the important connection of DIBs with solar system comets and its implied cosmological implications!**

XI. The Infrared Radiation Emmission (UIR) (Clue #11)

The paraffin assembly surrounding the fluorescent DIB carrier (MgTBP) essentially prevents emergence of its fluorescence emission radiation. However, it does permit the resulting long wavelength infra-red signals to be emitted from their matrix environment. Each one of these infra-red vibrational frequencies corresponds to one of the vibrational frequencies of MgTBP (Johnson 2006 Table 1). This accounts for **all** of the UIR radiation (Peters 2013), including the enigmatic 20 cm^{-1} DIB (Műller *et al.* 2013). This ultra-low 20 cm^{-1} vibrational frequency displays hyperfine substructure (hfs), resulting from the central Mg atom in the MgTBP molecule vibrating perpendicular to its molecular plane. 10% of Mg isotopes have a magnetic moment $\boldsymbol{\mu}$ and a spin of (**I=5/2**) which would produce a splitting of (**2I+1**) states in a magnetic field **H**, with energy $\boldsymbol{\mu} \cdot \mathbf{H}$. Műller's *et al.* (2013) published spectra does indeed display hfs of **6** (as theory predicts) individual substructures within its 20 cm^{-1} DIB.

XII. The Anomalous Star HD 204827

134 DIBs toward sightline HD 204827 (Hobbs *et al.* 2008) display a set of DIBs which differ significantly from DIBs toward HD 183143. These anomalous, so called C_2 DIBs result from exposure to an intense UV radiation field, which causes some partial as well as some complete disruption to its DIB carrier as well as to its immediate paraffin enclosures. DIBs in Table 14 illustrate the increase of its

fragmented array of DIB splittings, where none of the pairs in this table reveal the customary long-ranging list of vibrational frequencies (such as were noted in sightline towards HD 183143). The broad 4881 Å DIB is surprisingly missing in the 134 DIB listing.

XIII. Discussion: Origin and Evolution of Porphyrins

The abiotic synthesis of porphyrins is addressed by Gordon Hodgson (1972, 1973) and by Johnson *et al.* 1973. The DIB identification of the MgTBP molecule in two separate comet fluorescence signals (see preceding Blue DIBs segment) provides an opportunity to view this discovery on the wider scope of Cosmochemistry and Astrobiology. Since DIB signals are observed in other galaxies (Welty 2014; Cox *et al.* 2006, 2007; Cordiner *et al.* 2008, 2011; Smith *et al.* 2013) this would imply that comets and by extension solar systems (Johnson 2015) are present throughout the universe. **Furthermore, this abiotic porphyrin molecule MgTBP exposed to heat and water forms proto porphyrin IX (Johnson 1970), which is an important pre-life precursor molecule.**

XIV.

The two year (1965-1967) sustained search for the elusive DIB carrier MgTBP (Tetrabenzoporphyrines)

The financial support from a NASA contract provided the means to support two full time technicians, an organic chemist and a part-time secretary. The complete content of all published spectroscopy literature was scanned daily (obtained from the CalTech library, by the hired secretary). Hundreds of compounds were systematically studied/ scanned spectroscopically, including a large variety of "polyaromatic compounds (PAHs). Three of the dominant DIB spectral bands (4428Å, 6284Å and about 5780Å) had to be matched. Thus, PAHs could readily be eliminated from the list. Finally, as a result of this systematic search program (searching for the desired spectral match), we concentrated on the TBPs:(MgTBP, FeTBP and the freebase variety, TBP. The chemist, Dr. Castro, synthesized all of them.

The complete study and all experimental research are documented in a **NASA Report", F.M.Johnson, "Spectroscopic studies of Interstellar grains, 1970 CR-1667**

Postscript: 1. These reports were distributed to NASA Ames, where PAHs were continued to be studied and publicized as potential, yet, **unsuccessful** DIB candidates for about 6 decades.

2. To NASA Goddard, where Bert Donn unsuccessfully attempted to duplicate MgTBP spectra. It was a failed plagiarizing attempt.

The key DIB discoveries of MgTBP: (see also Chapt.6)

1. **It is an ultra-stable fluorescent molecule, situated within a paraffin "Schpolski" matrix**
2. **Its fluorescent signals are trapped within the matrix, hence, transmitting mainly absorption bands.** It has fooled astronomers for about a century.
3. **MgTBP is subjected to a variety of different electric fields ("Stark Effects") which split its primary electronic states by different amounts.** Hence, a large number of experiments are needed in order to obtain some of the variables.
4. **MgTBP has about 200 vibrational frequencies. Specific vibrational frequencies can be shown to corroborate almost all of the DIB transition frequencies.**

XV. Conclusions

The 11-step process outlined in this review is based on the many unique properties of the identified "DIB carrier" molecule (MgTBP) which include:

a) Extreme thermodynamic stability (it sublimes at 500 C). It has over 200 potential vibrational frequencies by which it can oscillate (which also accounts for its stability).
b) It is enveloped by a loose-fitting paraffin structure, which minimizes potential structural damage to its precious cargo.
c) The harsh UV interstellar environment requires DIB carrier protection with some type of shielding environment. Bare molecules are subject to UV degradation. Among the 170 discovered molecular species in the interstellar medium, there are at least 19 fragments which had once been part of the MgTBP molecule.
d) MgTBP's dominant (experimentally confirmed) 4428Å (Soret) absorption bend has a molecular extinction coefficient of 1.5×10^6. Its fluorescence (upon blue excitation) has a 90% conversion efficiency, which implies that a very small quantity of these molecules makes their presence known in the ISM. The 4428 Å DIB bandwidth also matches the lab data (Johnson 2006).
e) MgTBP has two excited states (degenerate electronic energy levels). Each one of them has a superimposed set of vibrational energy levels. This facilitates possible absorption into each of the excited states and their subsequent fluorescence.

f) Its degenerate energy levels are split by various amounts due to external electric fields, (Stark Effect), which result in 24 different origins, where each one of them has a superimposed set of vibrational energy levels. MgTBP has a potential of about 200 vibrational energy states. The totality of these possibilities allows for the production of over 500 vibronic absorption and fluorescence signals.

g) Fluorescent signals, except for 6614Å, are suppressed by various combinations of paraffin protectors (cages). However, its generated IR signals are being emitted and received on earth as UIR emissions.

Question: Why only this single MgTBP molecule? Why not some other complex molecule (responsible for DIBs)?

1) **Answer**: Its large molecular extinction coefficient (for the dominant DIBs) makes it stand out relative to other potential molecules which have far smaller oscillator strengths (f) values. Other molecules could exist but would probably fall far below present detection limits. MgTBP is a very efficient fluorescent emitter, most others are not. It's unique thermodynamic stability, is absent in most other molecules. Hence, there is an additional advantage related to its survival in its unique plasma jet "production phase", (due to its thermodynamic stability- not present in other molecules).

2) Atomic abundance requirements in the ISM would rule out molecules with exotic atomic species in their structure.

3) **The a priori prediction criteria**: Johnson (1972) published 30 experimental absorption and fluorescence spectra. Each one of them corresponds to a DIB (Hobbs *et al.* 2009). Of the 30 experimental DIBs, only 6 were known in 1971. Thus, **24** of them satisfy the a- priori prediction criteria.

4) All UIR signals are accounted for by vibrational frequencies of MgTBP. Tables 5-12 are a listing of DIBS as a function of electronic pair splitting, and each individual DIB is accounted for by a known MgTBP vibrational frequency.

A Final Thought (Relevance)

It should be universally recognized that life's evolutionary development depended on an effective utilization of incoming solar energy (see Chapt.6). Furthermore, such utilizations depended on the availability of porphyrins (a distinct variety of chlorophylls). MgTBP is an abiotic, precursor, porphyrin molecule.

References

A'Hearn, M.F., D.D. Wellnitz, R. Meier, 2013, Proc. IAU Symp. **297**, 216 (Eds. Cami, Cox)

Bondar, A., and J. Krelowski, 2012, Proc. IAU Symp. **297**, 135 (Eds. Cami, Cox)

Clayton, G.C., 2013, Proc. IAU Symp. **297** (Eds. Cami, Cox)

Cochran, A.L., W.D. Cochran, 2002, Icarus **157,** 297

Cohen, M. *et al.*, 2004, A.J. **127**, 2362

Cordiner, M.A., et al, 2008, A & A **480**, L13

Cordiner, M.A., *et al.*, 2011, ApJ Letters **726**, 39

Cox, N.L.J., *et al.*, 2006, A & A **447**, 991

Cox, N.L.J., *et al.*, 2007, A & A **470**, 941

Donn, B., R.K. Khanna, 1980, Astrophy. Space Sci., **68,** 19

Duley, W.W., S. Kuzmin, 2010, ApJ Letters 712**,** 165

Ehrenberg, B., F.M. Johnson, 1990, Spectrochim. Acta A **56** (10), 1521

Even, U., J. Magen, J. Jortner, 1982, J. Chem Phys. **77**, 4384

Even, U., 1989, Private Communication

Fan, H., *et al.*, 2017, ApJ Letters **850**, 194

Fielding, P.E., A.W.H. Mau, 1976, Aust J. Chem. **29**, 933

Galazutdinov, *et al.*, 2008, PASP **120**, 178

Galazutdinov, *et al.*, 2016, MNRAS

Geballe, T.R., 2013, Proc. IAU Symp. No**. 297**, 64

Geballe, T.R., *et al.*, 2011, Nature **479,** 200

Gouterman 1961, J. Mol. Spectros. **6**, 138

Heger, M.L., 1922, Lick Obs. Bull, 10, No. **337**, 146

Herbig, G.H., 1998, ApJ Letters **331**, 999

Herbig, G.H., 1995, Ann. Rev. Astrophys. **33**, 19

Herzberg, G., 1935, Mem. R. Astrom. Soc., Liege 4, 15, 291

Herzberg, G., 1967, IAU Symp. 31, **91** (Woerden, H.P., Ed.)

Herzberg, G., 1971, Vistas in Astronomy 2, **415** (Beer, A., Ed.

Hobbs, L.M., D.E. York, T.P. Snow, *et al.*, 2008, ApJ Letters **680**, 1256

Hobbs, L.M., D.E. York, J.A. Thorburn, *et al.,* 2009, ApJ Letters **705**, 32

Hodgson, G.W., 1972, N.Y. Acad. Sci. **194**, 86 (F.M. Johnson, Ed. "Interstellar Molecules and Cosmochemistry")

Hodgson, G.W., 1973, N.Y. Acad. Sci. **206**. 670

Hodgson, G.W., B.L. Baker, 1967, Nature **216**, 29

Hodgson, G.W., B.L. Baker, 1969, Geochim. Cosmochim. **33**, 943

Irvine, William, M., M. Okishi, N. Kaifu, 1991, Icarus 91, **2**

Jiang, B.W., *et al.,* 2013, Proc. IAU Symp. **297** (Eds. Cami, Cox)

Johnson, F.M., 1955, Columbia University Radiation Lab Progress Report, 30

Johnson, F.M., 1967, Use of Space Systems for Planetary Geology and Geophysics, **17**, 51

Johnson, F.M., 1970, Spectroscopic Studies of Interstellar Grains, NASA CR- 1667

Johnson, F.M., 1972, Ann. N.Y. Acad. Sci. **187**, 186

Johnson, F.M., 1972, Mem. Soc. Roy. Liege 6 (III), 391

Johnson, F.M., D.T. Bailey, P.A. Wegner, 1973, Interstellar Dust and Related Topics IAU, Symp. **52**, 317-321

Johnson, F.M., 1991, Bull. Am. Astron. Soc. 23 (2), 933

Johnson, F.M., 1993, Bull. Am. Astron. Soc. 25 (2), 820

Johnson, F.M., 1994, NASA Conference Publication 10144, Tielens, 47

Johnson, F.M., 1995, *The Diffuse Interstellar Bands*, Kluwer Academic Publishers, Tielens, Snow, 341

Johnson, F.M., 2006, Spectrochimica Acta Part A **65**, 1154-1179

Johnson, F.M., 2015. Voyage Into Astronomy 2nd Edition, Lawyers & Judges Publishing Company, Inc.

Keith, T.S., *et al.,* 2013, Proc. IAU Symp. **297** (Eds. Cami, Cox)

Krelowski, J., 2013, Proc. IAU Symp. **297** (Eds. Cami, Cox)

Krelowski, J., *et. al.*, 1997, PASP **109**, 1135

Merrill, P.W., 1934, PASP **46**, 206

Merrill, P.W., 1936 ApJ **83**, 126

Merrill, P.W., O.C. Wilson, 1938, ApJ **87**, 9

Műller, H.S., *et al.,* 2013, Proc. IAU Symp. **297**, 197 (Eds. Cami, Cox)

Nandy, K., 1964, Publication Roy. Obs. Edinburgh **3**, No. 6

Personov, R.I., 1962, Optika I Spektroskopiya 15, 61

Shkirman, S.F., K.N. Solovev, 1965, Investia Akad. Nauk SSSR 29, 1385 (also Dokl. 10, 1965, 349)

Peters, E., 2013, Proc. IAU **297**, 187 (Eds. Cami, Cox)

Platenkamp, R.J., G.W. Canters, 1981, J. Phys. Chem, **85**, 56.

Rossing, T.D., N.H. Fletcher, 2004, *Principles of Vibration and Sound* (2nd Ed. Springer N.Y.)

Russelle, R.W., *et al.*, 1978, ApJ **220**, 568

Scarrott, S.M., *et al.*, 1991, MNRAS, **252**, 229

Scarrott, S.M., *et al.*, 1992, MNRAS, **255**, 11

Schmidt, G.D., *et al.*, 1980, ApJ Letters **239**, L133

Sevchenko, A.N., *et al.,* 1967, Sov. Phys.—Dokl. 11, 587

Sevchenko, A.N., K.N. Solov'ev, A.T. Gradyushko, S.F. Shkirman, 1967, Dokl. Arad. Nauk SSSR, **587**

Sevchenko, A.N., K.N. Solev'ev, S.F. Shkirman, Kachura, 1965, Dokl. Arad. Nauk SSSR **10,** 149

Shpolskii, E.V., 1962, Soviet Phys.—Usp. 5, 552 (a)

Shpolskii, E.V., 1962, Usp. Fiz. Nauk 77, 32 (b)

Shpolskii, E.V., *et al.*, 1952, SSSR 87, 935

Smith, K.T., *et al.*, 2013, Proc. IAU Symp. **297**

Smith, Wm.H., 1995, The Diffuse Interstellar Bands, **3** (Tielens and Snow Eds.)

Snow, T.P., 1995, The Diffuse Interstellar Bands, 325 (Tielens and Snow Eds.)

Snow, T.P.,2013, Proc.IAU Symp. **297**

Underhill, A.B. and G.A.H. Walker, 1966, MNRAS **131**, 475

Van Winckel, H.V., 2013, Proc. IAU Symp. **297** (Eds. Cami, Cox)

Van Winckel, *et al.*, 2002, A & A 390. 147

Welty, D.E., *et al.*, 2014, ApJ Letters **792**, 106

Whitford, A.E., 1958, Astrom. J. **63,** 201

Whittet, D.C.B., *et al.*, 2004, ApJ Letters

Williams, T.B., P.J. Sarre, K. Spekkens, R.K. de Naray, C.M. Marshall, 2015, SALT Science Conference, SSC 2015, **045**

York, B., *et al.*, 2013, Proc. IAU Symp. **297** (Eds. Cami, Cox)

York, Donald G., 1971, ApJ Letters **166**, 65

Zaleskii, J.E., *et al.*, 1973, Dokl. Akad. Nauk SSSR **18**, 320

Zaleskii, J.E., *et al.*, 1975, Dokl. Akad. Nauk SSSR **19**, 589

Chapter 9

Quasars

(Quasars are infant galaxies in their earliest development stages)

Synopsis

The Doppler Effect

We must first discuss a slight technicality in order to understand the topic of quasars. By what method can we determine the velocities of objects which emit signals at specific frequencies, acoustic electromagnetic?

As the object moves towards an observer, the number of waves per second (the frequency) changes. In general, the frequency of a wave is the number of crests that we count as the wave passes by. If the source that emits the wave is coming towards us, we would be counting more of these crests, or an increased frequency. If the source is moving away from us, we would be counting less. For example: we have all heard the fire engine bell ringing as it approaches us. Have you noticed the change in pitch as it first approaches and then recedes away from us? The pitch or tone changed thus: It was at a higher frequency as it approached and then at a lower frequency as it moved away. This is a vivid demonstration of the well-known Doppler Effect with sound waves. The effect works equally well with either light or sound waves.

If the source is stationary and we were to be moving towards the source, the frequency would also increase. It is the relative velocity between source and receiver that counts. The relative line of sight velocity between any emitter and receiver can be determined by means of this Doppler Effect.

Recall the relationship between wavelength and frequency, wavelength × frequency = speed of wave or $\lambda \bullet v = c$. Imagine that we have a hot hydrogen source with terribly excited atoms which emit their characteristic photons. A short distance away, we have set up a spectroscope that examines these photons as they make their way through the instrument. We shall note that the wavelength of one specific emission line is λ_0. We could equally well use any of the lines of hydrogen. However, once we have picked out the line of λ_0, we shall ignore all others.

Now, as the excited source (with its continuously emitting characteristic photons) moves toward the observer, the waves seem to be pushed together. We will be counting more of these waves per second. The line we were studying is now no longer λ_0, but is shifted towards shorter wavelengths (blue shift) by an amount $\Delta\lambda$.

Let us repeat the same experiment, except this time, the source is made to recede from us. We would be counting less wave crests per second. The resulting wavelength has been lengthened by an amount $\Delta\lambda$ towards the red portion of the light spectrum. When scientists (usually astronomers) talk about a red shift, they are referring to this very phenomenon.

Now, put these qualitative concepts on a quantitative footing. We shall define a quantity z as the ratio of a Doppler shift ($\Delta\lambda$)to the" un-shifted" wavelength λ_0, i.e., $z = \Delta\lambda/\lambda_0$.

The speed of the object (provided it is much slower than c) can be directly calculated once z is known. (Use equation: $v = cz$ where c = speed of light.) If the object's velocity is almost as large as c, then a special relativistic formula is needed.

In the special case of low velocity ($v << c$), the relativistic formula reduces to the usual formula for the Doppler Effect. Note that the Doppler shift is directly proportional (at the lower speed) to the relative velocity between the object

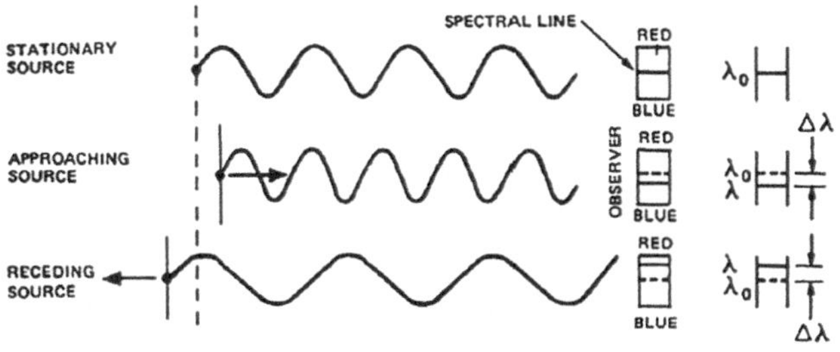

Figure 9.1. *Illustration of the Doppler Effect for both approaching and receding sources.*

and the observer ($v=c \bullet z=c\Delta\lambda/\lambda_0$). This is a very important physical phenomenon. It is one of the keys by which astronomers have unlocked some of nature's secrets. It enabled them to determine the relative speeds of stars with respect to the earth by merely observing the Doppler shift in the spectral lines. The sun's speed around the nucleus of our galaxy is also measured with the aid of the Doppler Effect.

Is it possible to construct a believable physical model of an object that would simulate the observed red shift? These are some of the questions that we shall confront as we begin our overview of the enigmatic Quasars. However, there is one technicality that requires a definition:

The Hubble Relationship

The recession speed (v) of distant galaxies (determines by the Doppler Effect) is directly proportioned to its distance (d) from Earth.

$$V=H_0\times d$$

Where v is measured in km/sec

d is measured in units of megaparsecs (1Mps = 3.09×10^{19} km)

H_0 is Hubble's constant = (about) 71 km/sec per Mps

Implications: The more distant a galaxy is from our planet, the faster it is moving away from us Conversely, if the speed of a galaxy is measured spectroscopically, its distance from us can theoretically be established, provided that one is certain in not having mistakenly overlooked the possibility of a revolving system around its central object. Extensive studies of galaxies in our universe (see Chapter 10) have so far never yet found a non-revolving galactic structure.

The study of Quasars is our next topic, subjected to a forensic analysis.

Quasars most likely exhibit the very beginning stages of a galaxy's early development, with rudimentary orbiting stars and accompanying gaseous clouds.

With a z here and a z there.
The quasars are just everywhere.
Assistants, assistants, what is their distance?
Excite them not, no matter what!
Be ready for the Red Shift plot. (FMJ)

Introduction to Quasars

Background:

The discovery of quasars in 1963 generated a great deal of excitement.

Let us first briefly reconstruct the state of astronomy at that time: A number of large radio telescopes were in operation. Feeble radio emission from the Sun was detected. Our galaxy was being mapped by means of its 21-centimeter radiation. Radio emission at various frequencies was detected throughout the galaxy and being mapped. Two distinct radio sources were located and identified. They were small in size, as could be ascertained when they were occulted (eclipsed) by the moon. Their position was eventually determined by means of interferometric techniques.

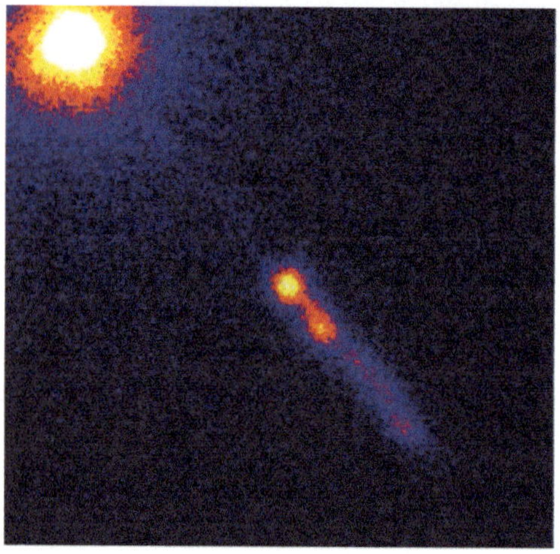

Credit: NASA/CXC/ SAO/H. Marshall et al., X-Ray Spectra of Quasar 3C 273

By 1963, there were four such "radio stars." The accurate location in the sky of radio source (3C 48) happened to coincide with a fairly faint star like object, using existing survey photographs. (The designation of a" radio star" such as (3C48), refers to Cambridge's third catalogue, indicating that the 250-foot dish at Jodrell Bank near Cambridge was used in the search).

The strongest of the four radio sources (3C 273), provided the surprising result. Maarten Schmidt (1963 & 1964) took optical spectra of this radio source and discovered that it contained four emission lines, readily identified as the well-known hydrogen Balmer lines. However, the lines were shifted far to the red end of the spectrum. In fact, a shift of such magnitude had not ever been seen in other stellar spectra. Based on this apparent large redshift, it was concluded that 3C 273, and all quasars discovered since then, exist at extreme cosmological distances. This premature conclusion was based entirely on the Hubble Relationship, which relates the speeds of receding quasars with their distances. (Speeds were determined by their Doppler red shift while ignoring the possibility of Doppler blue shifts).

The star-like quasars also had other peculiarities. They emitted strong ultraviolet radiation and had wisps of nebulosity surrounding them. In view of their star-like characteristics, they were renamed quasistellar objects (QSO).

Some of the early benchmark's discoveries:

A. The first red shift determination of 3C 273 in 1963.
B. The determination in 1965 of a red shift as large as $z=2$ (which, if it were due solely to a Doppler red shift, would imply an enormous distance).
C. Fluctuations in quasar energies in both radio and optical regimes.
D. Intensity fluctuations involving timespans of weeks or months. Hence, quasars had to be small in size, of the order of light months across (which would imply an exceedingly small galaxy).
E. In 1966 came the first report of absorption lines associated with QSOs.
F. 1968 the discovery of multiple redshift systems, i.e., a given quasar was found to have a number of different z's. Where $z=\Delta\lambda/\lambda$ and $\Delta\lambda$ = spectral line shift.

Ever since the discovery of quasars and their large potential Doppler red shift, astronomers have gloated on the prospect of having discovered the very beginnings of our early universe. However, in order to arrive at these erroneous conclusions, it was required to completely **disregard any potential Doppler blue shifts** which would be consistent with young early type stars revolving about a central massive object. Indeed, as a result of their spectroscopic Doppler misinterpretations, other **dilemmas** began to surface:

- Hawkins (2009) reports an inconsistency in the established cosmological quasar Doppler distances. He stated that there is a missing special relativity time dilation effect in the allegedly far more distant quasars, relative to the closer ones. This represents a serious challenge to the cosmological QSO redshift interpretation. This dilemma demands consideration, since the validity of special relativity is widely accepted.
- The quasar distance anomaly was also analyzed by Geoffrey and Margaret Burbidge (1967), who noticed problems regarding the incredibly large and compact postulated energy source necessary to account for all the quasars' large energy output, if they are indeed at remote cosmological distances.

A very talented observational astronomer, Dr. Halton (Chip) Arp, specializing in the photography of unique galaxies, recognized a number of unexplained and puzzling peculiarities: He discovered examples of nearby galactic clusters which had quasars in their midst having completely different red shifts than its immediate neighbors. A large set of similar examples were known as discordant red shifts, (while astronomers insisted that there were all accidental coincidences). Arp's numerous discoveries of discordant quasar redshifts demonstrated galactic development scenarios. These were ignored by the astronomical community. In order to curtail further unpleasant discoveries by Arp, he was restricted from receiving telescope time. Arp subsequently left for Europe to continue his research. His life's work on this topic *Quasars, Redshifts, and Controversies* was published in 1987. The essence of the dilemma, however, has always been the absence of acceptable alternatives. This is partly why the cosmological Doppler shift QSO interpretation was so widely accepted. Incredibly as it will turn out, there are indeed reasonable alternatives.

Whenever an object rotates, the side receding away from you will create a spectral redshift, while the side coming toward you, will demonstrate a spectral blue shift. Thus, any rotating object should provide both of these values. The necessary equations for calculating frequencies and rotational velocity (v) values are as follows:

Redshift: $$f = \frac{f_0}{1 + v/c}$$

Blueshift: $$f = \frac{f_0}{1 - v/c}$$

Where:
f_0 = an intrinsic signal frequency
f = measured frequency
c = speed of light (2.998×10^5 km/s)
(Assuming there is a local oscillator of frequency f_0 situated on a rotating object.)

The time is ripe for a reexamination of Arp's discovered anomalies. In view of an improved appreciation of solar system developments, discussed in this text, one may well wonder whether galaxies should not have undergone similar evolutionary processes, but on a much larger and far grander scale, which should include considerations of nuclear physics.

Credits: X-ray: NASA/CXC/CfA/E. O'Sullivan Optical: Canada-France-Hawaii-Telescope/Coelum - NASA Image of the Day Gallery

Stephan's Quintet, a compact group of galaxies located about 280 million light years from Earth. This is one of the examples provided by Chip Arp which shows a direct link of a quasar associated with normal galaxies. The implication being that the quasar is located at a distance comparable to the other galaxies.

Here is a summary of enigmas associated with Quasars:

1. The further the distance of a quasar from us, the greater is the amount of energy it emits. (Even a high school physics student should find this is contrary to common sense).
2. It violates one of the results associated with special relativity (there is a missing time dilation effect associated with (allegedly) distant Quasars). This was the 2009 Hawkins discovery, previously mentioned.
3. If Quasars are alleged to be at extreme, far out cosmological distances, the source for their required energy output defies any nuclear mechanism known to science, exceeding even complete mass annihilation and converting it to energy!!
4. The present theory relies entirely on a few, selected spectral lines. interpreted exclusively as velocity of recession (redshifted) and ignores:
   - Possibility of rotation and revolution
   - Presence of blueshifted spectra indicative of rotation
   - Multiplicity of absorption spectra indicative of Quasars being engulfed in gas clouds
5. Assumes (unreasonably) an absence of galactic revolution about a central core. (All galaxies rotate and revolve around a central core. Why should quasars behave differently?
6. The spectra of quasar 3C 273 exhibits a Lyman-Alpha line in absorption with a z value of **zero**. I would like to see someone try to explain this as representing a far distant object!

Quasar 3C 273: A Novel Energy Mechanism and Spectral Interpretation Distance of Quasar 3C 273

The quasar 3C 273 is a unique example among a large array of enigmatic quasars. It is the brightest and has a partially decipherable spectroscopic signature (the Hydrogen Balmer Lines). Since 1963, it has received the greatest scientific attention from among the thousands of quasars, now being catalogued (Bahcall et al. 1996, Burbidge et al. 1964, Courvoisier & Robson 1990, Hazard et al. 2017, Kriss et al. 1999, Montigny et al. 1997, Sembach et al. 2001, O'Brien et al. 1989, Dietrich et al. 1999, Bahcall et al. 1991). Thus, the quasar 3C 273 will be re-examined as a typical prototype for all other quasars.

The spectral properties of early type supergiant O-type and Wolf-Rayet stars (Underhill 1966, P. 213), as well as Novae (Stratton & Manning 1939, Bowen & Edlen 1939) have an astounding resemblance to those of quasar spectra (Wampler & Oke 1967). The deduced quasar gas temperature of about 30,000 K (Ulrich 1980) is also comparable to those of the O-type supergiants. This suggests that quasars need not necessarily possess the peculiar and highly irregular characteristics now commonly ascribed to them. Based on our knowledge of these O-type supergiant stars, and the study of a large selection of galaxies, it is not unreasonable to postulate that quasars are a class of very young galaxies in their earliest development phases.

The distance of quasar 3C 273 is calculated using the standard equation:

$$M = m + 5 - 5\log(d) \qquad (1)$$

Where the absolute magnitude (M) is in the range of −8 to −10 (based on supergiant O-type stars), the apparent magnitude (m) is equal to 12.9, and therefore the distance (d) would have a minimum value of 158 kpc and a possible maximum value of 400 kpc. These distances correspond to other well-known, nearby galactic systems, such as NGC 6822, NGC 147, and NGC 185 (Motz & Duveen (1966), Pg. 584).

At the (erroneously) assumed cosmological distance of 900 Mpc, the total luminosity for quasar 3C 273 (corresponding to $z = 0.158$) is given as 2.2×10^{39} Watts, and its radius is calculated to be 10^{16} cm (Ulrich (1980)). A re-evaluation of this luminosity, applicable to a (more likely) distance of about 200 kpc, would correspond to a luminosity (L) of 1.1×10^{32} Watts, a value comparable to O-type stars in our own galaxy. The Stefan-Boltzmann relationship and a gas temperature of 30,000 K produce a black-body luminosity per unit area (Q) of 4.593×10^{6} Watts cm^{-2}. Substituting this luminosity value into the following equation:

$$L = QA \qquad (2)$$

Where A is the total surface area. A (realistic) radius of 2.63×10^{11} cm is deduced, which is only 3.7 times larger than the current radius of the Sun.

Let us compare these results based on an analysis in Chapter 1. The stellar mass-luminosity relationship is given by the equation $y = 0.6318e^{0.1236x}$ where y is the logarithm of the luminosity, and x is the logarithm of the mass. The predicted masses range from 10^{32} kg to 1.4×10^{32} kg, or roughly 100 solar masses. These are very reasonable numbers, based on a statistical analysis of a large number of early type stars.

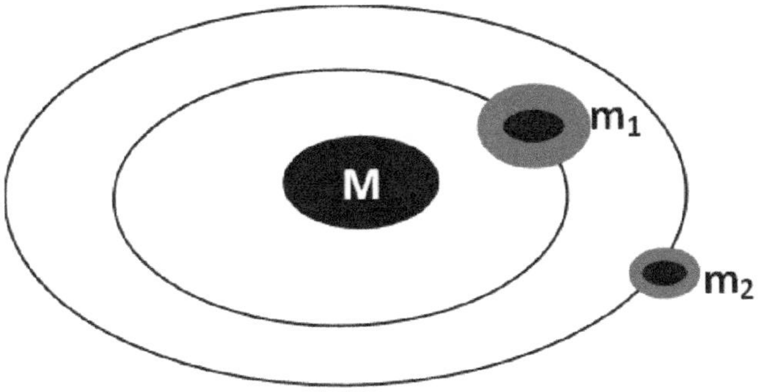

Figure 9.2 *A supermassive, dense nuclear fission core surrounded by orbiting stellar solar systems.*

Dynamical Motion Details

A paradox associated with almost all QSO research is an absence of QSO-specific intrinsic physical or dynamic orbital motion considerations. However, Hong-qiang et al.(2016) were able to extract three distinct periods (2.06, 13.03, and 21.15 years) within 3C 273. Their method included Fourier transform analysis, from an extended list of light intensity measurements, accumulated for nearly a hundred years. These results were incorporated in the development of a new quasar model.

Based on our knowledge of astronomical objects in our solar system and in our galaxy as a whole, one should not assume that a QSO would be devoid of angular momentum. The novel interpretation presented here discusses light variation of quasar 3C 273, as an occultation phenomenon caused by massive star-like bodies plus its planetary compliment revolving around a large, massive central core. Figure 1 shows a simplified dynamic configuration that is envisioned for quasar 3C 273. Based on a recent discovery that all planets and all stars contain identical nuclear power sources (Chapter 1, Johnson 2018), it is postulated that these properties also extend to regions beyond our own galaxy. Consequently, all developing galaxies are likewise expected to possess dense, centrally located nuclear fission-type spinning cores, which, as part of their development phase, eject O-type supergiant proto-stars (OTSs), possibly at relativistic speeds.

The bluish underlying background radiation, ubiquitous for all QSOs, can therefore also be attributed to nuclear fission reactor mechanisms. Consequently, quasars are merely proto galaxies (not an unfamiliar concept), which were ejected from super-dense galactic nuclei.

Quasar 3C 273 has both a small and a large-scale jet associated with it. Their properties were analyzed by Courvoisier (1998) and Marchenko et al. (2017). These jets are not being discussed except as to point out that, based on the above theoretical development, these observed phenomena would involve ejection, rather than postulated inward flow effects. This is in direct contradiction to an (erroneous) Schwarzschild's Black Hole theory for quasar energy production.

Ulrich et al. (1980) reported the presence of a redshifted 2175 Å extinction maximum. This is a very important discovery. By comparing this result to similar ones seen in various intergalactic clouds, it can unambiguously be established that organic molecules, such as pyridine and paraffins (Johnson 2006), are constituents of the dust that surrounds quasar 3C 273. This in turn would be indirect evidence for multiple solar system developments associated with this quasar.

As shown in Figure 1, and based on the above quasar interpretation, 3C 273 is assumed to possess a heavy, central, massive core, M, surrounded by circulating sub-cores of masses m_1 and m_2 and possibly more. All three of these masses are evolving, undergoing development phases, producing proto-stars, and extruding gases (primarily hydrogen) and other atomic species, all of which are byproducts of their nuclear fission processes. O type supergiant stars' spectroscopic output is similar due to its high degree of ionization (Underhill 1966). The combination of both orbiting and sub-orbiting motion yields dynamic orbital motion effects which can be discerned and unraveled if correctly interpreted from their combined Doppler red and blue-shift spectral lines.

Table 9.1 is a list of typical QSOs with their absorption and emission z values. This table presents the best clues as to the mechanism and dynamics within each quasar.

The new Johnson hypothesis, however, requires recognition for the existence of blue-shifted spectral lines. Given the large potential pool of available ionized atomic spectral features erroneously either ignored or misinterpreted, it was possible for astronomers to assign QSO spectra as if all the spectral lines represent indeed only red-shifted lines exclusively. This is definitely not the case, as will be demonstrated with examples.

I therefore wish to draw attention to the presence of a compendium that lists over 100,000 spectrum lines between 10,000 A. and 2,000 A. arranged in order of decreasing wavelengths, which covers all available atomic elements, and their spectrum emission lines:

MIT Wavelength Tables

Measured and Compiled Under the Direction of George R. Harrison John Wiley & Sons, Inc. published in 1939

Consequently, it is readily possible to mistakenly assign any spectral feature to a random element listed in the catalogue as a Doppler redshift, instead of its correct, blue-shifted assignment.

Table 9.1
Selected QSOs
With Absorption lines in Their Spectra*

Object	Z_{em}	z_{abs}
NGC4151	0.00327	0.00233, 0.00143, 0.00047
BSO1	1.241	1.241
3C 205	1.534	1.538
Mk 132	1.758	1.7316, 1.3518, 1.2750
B194	1.864	1.8952, 1.8366
PHL 122	1.910	1.934
0119-04	1.955	1.965
PHL 938	1.955	1.906
PHL 5200	1.98	1.90–1.98, 1.9502, 1.8910
Ton 1530	2.046	2.0553, 1.9798, 1.9362
1331+170	2.082	1.7851, 1.7754, 1.3273
0237–23	2.223	2.2013, 2.1758, 1.9551, 1.6737, 1.6706, 1.6564, 1.5955, 1.5566, 1.5503, 1.5139, 1.4145, 1.3644
4C 25.05	2.360	2.3682, 2.3456, 2.3421, 2.3004, 2.2754, 2.2432
5C 2.56	2.390	2.367
PHL 957	2.69	2.6614, 2.5504, 2.5423, 2.3090, 2.2250
4C 5.34	2.877	2.8754, 2.8098, 2.7701, 2.5925, 2.4739, 2.1819, 1.7758
OH 471	3.398	3.343, 3.246, 3.191, 3.122, 2.911
OQ 172	3.53	3.094, 3.089, 3.066, 2.698, 2.691, 2.564

*Data from E.M. Burbidge and G.R. Burbidge, Ap. J. 202, 287 (1975).

The Spectroscopic Evidence and its Reevaluation

Several research articles that (incorrectly) discuss quasar spectra, assume that some spectral features arise from intergalactic material. However, they neglect to point out that there is no visible excitation mechanism to produce emission features in the intergalactic medium. Furthermore, some z values of the supposedly intergalactic material are not too different from quasars' z values. Thus, a majority of quasar-related spectral features are associated directly with quasars, not intergalactic.

The unusual width of broad (200 Å) and (50 Å) spectral bands in the 3C 273 quasar emission spectra were never addressed by Greenstein & Schmidt (1964), and in fact, about half of its spectrum was left uninterpreted! Attempted interpretations by Wampler & Oke (1967), and others, have so far been unconvincing in view of their introduction of a largely random array of possible (mostly iron) ionic species. Consequently, a detailed reevaluation of quasar's 3C 273 visible spectra was initiated.

Using the three derived periods for 3C 273 from Hongqiang et al. (2016), it was possible to extract two separate z values for the main orbits of both m_1 and m_2. Using the non-relativistic Doppler shift frequency (f) equation:

$$f = \frac{f_0}{1+z} \qquad (3)$$

This equation will be modified to include the possibility of evaluating the velocities (V) of its sub-orbiting stellar components. The z value would therefore correspond to the velocities of core masses m_1 and m_2. The derived V values would provide velocities of its orbiting stars, using the sum and difference of the total velocity components. Thus, Equation 3 is modified to:

$$f_1 = \frac{f_0}{1+z+V} \qquad (4)$$

$$f_2 = \frac{f_0}{1+z-V} \qquad (5)$$

Here are the detailed calculations:

The two Doppler equations 4 and 5 are solved simultaneously:

$$\left(1+z\right) = \frac{f_0\left(f_1+f_2\right)}{2f_1f_2} \qquad (6)$$

This result is then inserted into Equation 4 to elicit V value.

Wampler and Oke's (1967) spectroscopic measurements of two broad emission features centered about the $H\beta$ Balmer emission line provided the starting point for a detailed analysis of object m_1. The symmetry around the $H\beta$ line suggested that there must be some logical causation. It was hypothesized, therefore, that this would correspond to a sub-orbital revolution around the mass m_1, resulting in both red and blue shifted velocity values. Indeed, this was found to be the case. The center of each of the broad bands was measured from the published data (Wampler & Oke 1967) at 6081 Å and 5277 Å respectively. The $H\beta$ center line is at 4861 Å. Consequently, it was postulated that one should be able to extract both a new z value as well as a velocity (V_1) for the orbital motion of stars about its center m_1.

Substituting numerical values into Equations 6 and 4 gave: z=0.168 and V=0.082. In view of the extreme width (200 Å) of each of the bands, the above analysis was repeated using the extreme inner and outer boundaries of these two symmetric emission bands. The results were as follows:

For the set of innermost distances, z=0.165 and V=0.07. For the set of the outermost distant wavelength values, z=0.171 and V=0.09. Their averages are z=0.168 and V=0.08, which correspond to the numbers that were derived for the center of the band. These results indicate that there is indeed a range of velocities inherent in the set of orbiting stellar objects, as well as that of the central mass m_1.

Published UV spectra of quasar 3C 273 (Ulrich et al. 1980, Ulrich 1980) provided additional examples of broad emission bands centered about prominent spectral emission lines. There were two examples: the $Ly\alpha$ at (1216 Å) and C *IV*, which was measured at 1791Å, but in its rest frame is at 1546.6 Å. Again, using Equations 6 and 4, the derived values obtained were: $Ly\alpha$ z=0.143 and V=0.090; and for C *IV* z=0.148 and V=0.098.

Similar calculations were performed for the narrower 50 Å wide Balmer line quasar emission bands.

Orbital parameters for the mass m_2 were similarly derived. Best estimates are provided in Table 9.2 for an estimated wide range of orbital periods (either 2 or 32 years).

Thus, these broad bands would correspond to a range of orbits associated with a subgroup of stars circumnavigating about their sub-nuclear masses m_1 and m_2.

Assuming that masses m_1 and m_2 revolve in circular orbits around the central core M, under the influence of a standard gravitational field, the mass of M was calculated using the velocity and period of each object, m_1 and m_2, in the following equation:

$$M = \frac{aV^2}{G} \qquad (7)$$

Where a_1 and a_2 are the orbital radii of m_1 and m_2 respectively, and v_1 and v_2 are the deduced velocities of m_1 and m_2 respectively revolving about the centrally located massive nuclear object M (See Table 9.2).

The derived z values provided the velocities of each of the masses, m_1 and m_2, about the quasar's galaxy center, M. Whereas the V values correspond to the orbital velocities of stars/planets circumnavigating m_1 or m_2.

The 50 Å bandwidth, associated with all the 3C 273 Balmer lines, is interpreted to reflect a narrower band of stellar orbits navigating about its central mass m_2. The orbital evaluation of both m_1 and m_2 produced practically equal values for the heavier nuclear central core mass M. The deduced average value for the mass M is 1.5×10^{41} kg (Table 9.2).

Equation 7 was used a second time to obtain the mass of m_1 based on the orbital parameters around m_1. The results of these calculations are tabulated in Table 9.3.

(The calculated value for the m_1 mass is well within the expected range of values for its central stellar sub-core.)

The deduced *z* values for 3C 273 are clearly all different, yet fall into a relatively narrow range: 0.145, 0.168, 0.158. It is probably significant that the newly derived *z* values average out to 0.157.

The hydrogen Balmer lines observed in emission by Greenstein and Schmidt (1964) were clearly red-shifted Doppler lines. It was assumed, since the predicted, blue-shifted spectra were not seen in emission, that indeed this could only be explained by the cosmological Hubble expansion

Table 9.2
3C 273 Deduced Orbiting Parameters

	Z	Velocity (10^8 m/s) V	Period (Years)*	Period (10^7 s)	Deduced Orbit Radius (10^{15} m)	Deduced Mass of M (10^{41} kg)	Average Mass of M (10^{41} kg)
m_1	0.1678	0.503	13.03	41.1	3.3	1.3	<1.5>
m_2	0.1585	0.476	21.15	66.8	5.06	1.7	

*Hong-qiang et al. (2016)

Table 9.3
Orbiting Parameters of Stars and Planets of m_1 and m_2.

	z	Velocity (10^8 m/s) V	Period (Years)	Period (10^7 s)	Deduced Orbit Radius (m)	Deduced Mass of m_1 and m_2 (kg)
Stars and Planets of m_1	0.0824	0.247×10^8	2.06*	6.5	2.56×10^{15}	2.36×10^{39}
**Stars and planets of m_2	0.00514	1.546×10^6	2	6.5	2.46×10^{14}	0.57×10^{36}
**Stars and planets of m_2	0.00514	1.546×10^6	32	101	1.6×10^{13}	8.8×10^{36}

*Hong-qiang et al. (2016)

**Extrapolated based on incomplete data.

Table 9.4
Predicted Doppler-Blue-Shifted Balmer Lines For z=0.158

Balmer Lines	(Å)	Predicted Blue-Shifted Bands (Å)
Hα	6562.8	5526
Hβ	4861	4093
Hγ	4340.5	3655
Hδ	4101.7	3453

hypothesis. The unexpected discovery of Doppler blue shifted Balmer lines in quasar 3C 273 spectra is revolutionary. Surprisingly, these features are seen not in the expected emission, but in absorption. If one examines carefully the published spectra (Ulrich et al. 1980), three out of four of the blue shifted Balmer lines (Table 9.4). are discoverable in absorption in both the UV and visible region of the spectrum. These absorption features are not as impressive as the emission features, but they are clearly present. The quasar is oriented at an inclination of 60 degrees, surrounded by great quantities of gas. Due to these characteristics, its emission spectra would be obscured on its orbital return cycle, when emission lines have to pass through an extended gaseous atmosphere

This discovery confirms the theory outlined in this section.

A dense nuclear fission core is expected to produce hydrogen atoms profusely as a result of β-decay. It would be spinning at a very high rate of speed. Emission spectra emanating from the precise center of its spinning nuclear core would produce Doppler shifted spectra in directions perpendicular to our line of sight Thus, its Doppler shifts could essentially be almost zero. This is precisely what is observed in the UV spectra of a specific Lyman-alpha line, published by Ulrich (1980).

A large number of conundrums/anomalies were meticulously identified by Halton "Chip" Arp. These are now readily explicable with just one minor assumption: quasars are very compact, early-stage galactic rotating structures, yet small for size determinations. Their enormous redshifts (large z) are readily recognized; however, their corresponding blue shifts would generally end up in the more difficult, far-UV region, beyond 3800 Å, hence this effect was so readily missed. Yet, for small values of z, one could and should have been able to recognize blue shifts. (See Table 9.5)

Table 9.5
Evidence for Intrinsic Rotation and/or Orbiting Bodies in Commonly Used Quasar Spectra

$f=f_0/[1\pm(v/c)]$						
Present Accepted Element Designation	**Possible Blueshifted Line (Å)**	**Present Accepted Element Designation**	**Possible Redshifted Line (Å)**	**Deduced Blueshift Velocity (v/c)**	**Deduced Redshift Velocity (v/c)**	**z**
Hα 6563 Å						
He II	4685.6	O I	8446.5	0.286	0.287	0.287
S II	4076	S III	9069	0.379	0.380	0.380
C II	2326	He I	10830	0.65	0.646	0.646
Hβ 4861.3 Å						
H	3970	Ne II	5754.6	0.183	0.181	0.183
Ne V	3426	O I	6300.3	0.295	0.294	0.296
Ne V	3346	Fe X	6374.5	0.311	0.311	0.311
Si II	1263	O I	8446.5	0.74	0.737	0.737
Hγ 4340.46 Å						
Ne III	3967	Ar IV	4711	0.086	0.085	0.085
O II	3726	O III	4959	0.141	0.142	0.142
Mg II	2798	He I	5876	0.355	0.353	0.354
Lα 1215.67 Å						
He II	1085	C II	1335	0.107	0.098	0.098
O VI	1034	Si IV	1396.7	0.149	0.149	0.149
S VI	937	N IV	1486 (?)	0.23	0.22	0.223

Unfortunately, the visible spectra catalog of neutral and multi-ionized atomic species, is crowded with a large number of spectral line possibilities, making identification both difficult and ambiguous. There is, however, one important and specific atomic identifier: the 21 cm Hydrogen line. Observations of small z quasars' hydrogen 21 cm spectra in 14 QSOs was reported by Hutchings et al (1987). Indeed, there are at least three clear-cut QSO examples which display two distinctly separated 21 cm features, associated with each one of the three quasars.

Conclusion

There clearly is a great deal more information encoded in quasar spectra which has so far been either misdirected or misinterpreted and which could in the end provide additional details regarding quasars' orbital parameters and their evolution.

"The discovery of QSO's large redshifts misled nearly all astronomers to the conclusion that they must be at very large distances, and thus would be a new and powerful tool for cosmological investigation. Most astronomers would like to accept this interpretation" (Burbidge & Burbidge 1967). This concept has remained unchanged for about half a century. However, the evidence clearly demonstrates that this paradigm of cosmological distances is no longer tenable, at least for quasar 3C 273, and by inference, for all other quasars.

The untenable dilemma for the Hubble cosmological quasar model is the confirmed presence of multiple z's, plus both blue and red spectral Doppler shifts in a single quasar's visible spectrum. Consequently, quasars are not at the once hoped-for cosmological distances, but exist among the nearby galaxies!

A reevaluation of Greenstein and Schmidt's (1963, 1964) original spectra in 3C 273 demonstrated the potential presence of at least two orbiting sub-cores with attendant stars and planetary development systems. Using the Doppler principle, it was possible to deduce the orbital parameters of one (or possibly both) of these sub-cores and its surrounding proto-stars, which provided both red and blue-shifted velocity values in the visible spectrum. Based on putting the quasar 3C 273 at distances between 200 kpc and 400 kpc, very reasonable luminosity and mass values, consistent with known galactic parameters, were established. These values were correlated with the recent discovery of the energy mechanism of all stars in our galaxy.

The resulting values for quasar's (3C 273) mass, luminosity, and distance are far more realistic than the parameters assumed for quasars at cosmological distances. Other anomalies, such as velocities greater than the speed of light, Hawkins' missing special relativity effect (2009), and the incredibly high energy demands of quasars at cosmological distances, are now mute.

A highly active nuclear core, with off-shoots and fission reactor developments, would constitute a fraction of the observed violent emission processes including $X-rays$, and the rest of the variable features in the electromagnetic spectrum.

The most serious challenge to the present quasar paradigm is the existence of a Lyman-α absorption line with a z value of zero, in the spectra of quasar (C 273). This is inexplicable under the current paradigm.

Appendix

There exists a list of multi-use emission spectral lines associated with both AGN and quasar spectra (Cox 2000, Pg. 596, Table 24.6, IR-UV emission line features commonly observed in AGN). It can readily be shown, as Table 9.5 demonstrates, that a large number of these emission lines are in fact possibilities for both red and blue-shifted Doppler lines originating from the hydrogen spectrum. Table 9.5 lists 13 examples of z values which represent Doppler effects for both blue and red shifts. Note one specific example for the $H\beta$ reference frequency 4861.3 Å, where a z value of 0.183 is given. This may possibly be relevant for quasar 3C 273, since its spatial orientation angle of 60° is postulated:

$$(sin60)\ x\ 0.183 = 0.158 \qquad (8)$$

The Doppler data displayed in Table 9.5 represents values for "edge-on" spectra. These values clearly will be modified for different orbiting galaxies having random angular orientations.

References

Arp H., 1987, "QSOs, Redshifts, and Controversies". Interstellar Media

Arp H., Bahcall J. N., 1973. W.A. Benjamin, Inc. Bahcall J. N., Jannuzi B. T., Schneider D. P., 1991, ApJ, 377

Bahcall J. N., Kirhakos S., Saxe D. H., 1996, ApJ, 479 Bowen I. S., Edlen B., 1939, Nature, 143, 374 Burbidge G. R., Burbidge E. M., 1967, "Quasi-Stellar

Objects". W.H. Freeman and Company

Burbidge E. M., Burbidge G. R., Crampin, D. J. 1964, ApJ, 140

Burbidge E. M., Burbidge G. R. 1967, Freeman Co.

Courvoisier R. J., 1998, A&A Rev., 9, 1

Courvoisier T. J., 2001, Encyclopedia of Astronomy & Astrophysics,

Courvoisier T. J., Robson E. I. Blecha A., 1990, A&A, 234, 73

Cox A. N., 2000, "Allen's Astrophysical Quantities Fourth Edition". Springer-Verlag

Czerny B., Zbyszewska M., 1991, MNRAS, 249, 634 Davidson K., Netzer H., 1979, Reviews of Modern Physics, 51, 715

Dietrich M., Wagner S. J., Courvoisier T.-L., et al. 1999, A&A, 351, 31

Greenstein J. L., Schmidt M., 1964, ApJ, 140, 1 Hawkins M. R., 2009. pp 24–+

Hazard C., Jauncey D., Goss W. M., et al. 2017, arXiv, Hong-qiang Y., Hao-jing Z., Xiong Z., et al. 2016, Chinese Astronomy and Astrophysics, 40, 1

Johnson F. M., 2006, Spectrochimica Acta A, 65, 1154 Johnson F. M., 2015, "Voyage into Astronomy Second

Edition". Lawyers and Judges Publishing Company, Inc.

Johnson F. M., 2018, Planet. Space Sci., 151, 125 Kriss G. A., Davidsen A. F., Zheng W., et al. 1999, ApJ, 527, 683

Lee G., Kriss G. A., Davidsen A. F., et al. 1992, Testing the AGN Paradigm, ed. S.S. Holt, S.G. Neff, and C.M. Urry, pp 159–+

Marchenko V., Harris D. E., Ostrowski M., et al. 2017, arXiv,

Montigny C. V., Aller H., Aller M., et al. 1997, ApJ, 483, 161

Motz L., Duveen A., 1966, "Essentials of Astronomy". Wadsworth Publishing Company, Inc.

O'Brien P. T., Zheng W., Wilson R., 1989, MNRAS, 240, 741

Schmidt M., 1963, Nature, 197, 1040

Sembach K. R., Howk J. C., Savage B. D., et al 2001, ApJ, 121, 992

Stratton F. M., Manning W. H., 1939, "Atlas of Spectra of Nova Herculis 1934". Cambridge University Press Türler M., Paltani S., Courvoisier T. J., et al. 1999, A&AS, 134, 89

Ulrich M. H., 1980, Space Sci. Rev., 28, 89

Ulrich M. H., Boksenberg A., Bromage G., et al. 1980,

MNRAS, 192, 561

Underhill A. B., 1966, "The Early Type Stars, Vol. 6".

D. Reidel Publishing Company

Wampler E. J., Oke J. B., 1967, ApJ, 148, 695

Chapter 10

Supernovas and Pulsars

The Crab Nebula and its Pulsar

Credits: NASA, ESA, J. Hester and A. Loll (Arizona State University)

The Crab Nebula, as a typical example of a supernova remnant, is subjected here to a reexamination. Pulsars are core remnants which become visible subsequent to Supernova explosions.

The Sun's complex energy mechanism, according to the new Johnson theory, had provided the necessary input for unraveling several enigmas associated with a supernova's evolution. Moreover, it facilitated a large number of additional enigmas to be more readily explained, such as the causation for an onset of a supernova and its underlying substructures.

Historical Background

The recent discovery that all planets and stars operate on a complex fission energy mechanism (based on an exponential tight fitting mass-luminosity [Chapter 1]) had stimulated a reexamination of the Sun's overall operation (Chapter 2). This in turn, has raised to the forefront an analysis of our Sun's developmental life cycle and by extension all of the other stars in our galaxy. The new novel solar energy mechanism provided an opportunity to propose a different stellar development scenario; in fact, it is postulated that a rare stellar demise is entirely possible during a star's *early development* phase.

The enormous attention given to supernova studies by astronomers can be gaged by a 1982 review article by V. L. Trimble, which lists over 470 references in its bibliography. A supernova prototype and a well-researched example is the Crab Nebula, which resulted from an explosive event in 1054 C.E. (seen as a guest star by Chinese astronomers).

According to the new Johnson theory, stars commence their life cycle as compact rapidly spinning neutron cores, with initial core masses of approximately 50–96 times the Sun's current mass. The bulk of these core masses are spun off while simultaneously producing a clutch of diverse planets (Chapter 4 and Johnson 2015). A specific "defect" in a star's early development cycle will essentially terminate its normal evolution. It aborts its development by an explosive expulsion of its outer surface region, while simultaneously exposing its inner spinning neutron core.

The Crab Nebula has been the beneficiary of intense scientific observations and study, spanning almost a century. Here we shall reexamine this object and provide a new perspective based on a novel concept, using our Sun as a basis for both a normal and a hypothetical failed stellar development scenario.

The Crab Nebula and its Supernova History

The following facts associated with the Crab Nebula seem to be well established.

- It consists of a core object, and an expanding filament shell moving at a maximum speed of 1500 km/sec (Trimble 1968)
- There is an expanding synchrotron component with a similar expansion rate (Bietenholz et. al. 1991)
- Both components indicate an acceleration in their expansion at a rate of about 0.8×10^{-3} cm sec^{-2}
- The "core" of the prior progenitor star is now the **Crab Pulsar** which is moving with a tangential velocity of 123 km sec^{-1} along the major axis of the Crab Nebula towards the centroid of optical and x-ray emission and parallel to the **E** vector direction in the vicinity of the pulsar, where **E** is the electric field component (Wyckoff and Murray 1977).

- There is a proven kinematic connection between the pulsar and the Crab filaments (i.e. they both have a common origin (center).

Accordingly, since there were still many unknowns and unanswered questions about the Crab Nebula, a large number of assorted models were created. None of these models, however, provide satisfactory solutions to *all* the enigmas:

- What gave rise to the acceleration, presumably at its very beginning phase?
- What are the masses of the pulsar and the filaments?
- What provides the mechanism for x-ray production and the underlying origin of continuous electron production and the necessary magnetic field?
- What is the excitation mechanism for the light-emitting filaments of the Crab Nebula? Their distance is now orders of light years from the central pulsar. Translation: What could possibly provide the energy to excite molecules in the nebula that are light years distant from their origin?!

The solution to these dilemmas is based on the new Johnson theory: Thousands of highly active nuclear components (SNFRs), (the Q conglomerate) which once occupied the subsurface of the progenitor star and were ejected during the supernova event, now constitute the underlying structure of these filaments and gas clouds. SNFRs have a relatively long half-life and are emitting neutrons, protons, electron, alpha particles, etc. which perform the function of exciting the surrounding gas. The nuclear core of a supernova is in fact, the standard core found in the center of every star, including our Sun. This core remains in place, and it is now still active in the Crab nebula's case, which is periodically emitting components known as SNFRs. SNFRs. get launched from the spinning core of the Crab nebula at about 10% the speed of light, a speed which is probably comparable to the surface rotation velocity of the Crab nebula's negatively charged core. Other electrically charged spinning objects are orbiting the core and subjected to the core's very high magnetic field. Such objects produce their own magnetic field and produce the observed electromagnetic pulses as electric dipole radiation.

Stellar stability (for all stars) is critically dependent on the equilibrium condition of Q's electrostatic repulsion being equal to its gravitational core attraction. Both effects obey an inverse square distance relationship. Overall stellar equilibrium is therefore very much electric charge dependent. Hence there are the following two possibilities:

a. A (sudden) diminished total charge of either Q or its nuclear core would result in a smaller size stellar radius, eventually leading to a *dwarf* star.

b. A sudden extremely large electric charge increase in the nuclear core's total charge could be catastrophic, or more likely, a ***sudden decrease*** in the gravitational attraction between the nuclear core and the Q conglomerate would result in Q's explosive (electrostatic) expulsion. If the total electric charge on the central core remains constant, while a mass M_2 gets spun off, then this rare event becomes a unique supernova.

Using the Crab nebula of 1054 C.E. as a typical example of a supernova, the following list of astronomical observations are explicable in terms of the new solar fission model (SFM) (Chapt.2).

- The star's central negatively charged spinning nuclear core becomes visible subsequent to the explosion. It is presently known as a neutron star, a pulsar.
- The Crab Nebula experienced an acceleration (at the time of the explosion), which was already noted by Baade in 1942, who obtained a value of $1.1 \pm .04 \times 10^{-3}$ cm sec^{-2} (a modern value is 0.8×10^{-3} cm sec^{-2}). The size of the present day filament structure is enlarged as a result of this acceleration, which is superimposed on its uniform expansion velocity.
An attempted solution, (F. M. Johnson and C.H. Townes, 1957), is herewith modified: The causation is a "Q ejection phenomenon" (outlined in detail below) which is governed by repulsive electric charges between the nuclear core and the negatively charged SNFRs, comprising the Q conglomerate.
- There is polarized synchrotron radiation caused by electrons revolving at relativistic speeds in a magnetic field estimated at 10^{-3}gauss, according to an early report by Mayall and Oort (1942). The origin of this magnetic field and the relatively short half-life of the" optical" electrons, had been one of the Crab nebula's enigmas for decades. However, according to the new F.M. Johnson hypothesis, a multitude of ejected SNFRs produce the electrons, and since each SNFR is electrically charged and spinning, it also produces a magnetic field. The continuous nuclear activity of each SNFR results in its profuse emission of neutrons, electrons, nuclear particles and gamma rays. The total ejecta from this conglomerate are now observed and being recorded by a range of astronomical instruments. Thus, the intense x-rays and gamma rays from the nebula are due to SNFRs' immense nuclear activity.

- Crab Nebula's nuclear core has not ceased its production of individual SNFRs. Activity near the core is frequently observed: as far back as 1942, Baade had already noted fast light ripples (more recently renamed Scargle wisps [Scargle 1969]) moving from the core at speeds estimated at about 1/10 the speed of light. Dramatic images of these fast moving wisps were captured by a series of NASA's Hubble Space Telescope observations (J. Hester and P. Scowen 1996). Oort and Walraven (1956) were among the early astronomers who had already noticed the many anomalies of the Crab Nebula, whose solutions are finally being aired in this chapter.

The nebula's new initial development scenario also implies that the catastrophic event occurred during the early stages of the star's evolution and not towards the end of its life.

Scenario of a Supernova event:
The Crab Nebula and a more detailed Pulsar Analysis

The following calculations are an attempt at a more complete solution of the 1054 supernova catastrophe.

Measured Data:

Core recoil velocity = 123 km/s (Wyckoff and Murray 1977)

Maximum filament velocity = 1500 km/s (Trimble 1968)

Distance to Crab = 2 kpc

Present size radius of Crab Nebula filaments ≈ 1.7 pc

Pulsar period = 0.0334033 sec (Bejger and Haensel 2003)

Pulsar slow down = 4.20996×10^{-13} sec • sec^{-1}

Acceleration of ejector = 0.82×10^{-3} cm sec^{-2}

Velocity of Scargle wisps ≈ 0.1 c (Johnson theory: SNFR ejection from the spinning core)

c = speed of light.

A provisional numerical analysis of the Crab Nebula is presented here.

Assumptions:

- Assume mass of progenitor star (prior to explosions) = M_0 (at its early developmental phase) has established a nuclear fission complex (Q) at its subsurface with mass μ. The surface charge on μ is Q_1 and on the spinning nuclear core (M_3) is Q_2.
- Total Mass of filaments = μ
- Assume the supernova was caused by a sudden net central charge increase resulting from the spinning core mass (M_3) ejecting some of its mass, where M_0 is equal to the mass of the progenitor star, M_1 is the mass of the pulsar, M_2 is the mass of the ejected object. ($M_1 + M_2 = M_3$)
- An explosion resulted as the progenitor star's subsurface, (of mass μ) was ejected. (This ultimately became the Crab Nebula).

$$M_3 + \mu = M_0 \quad \text{(1)}$$

At equilibrium, the gravitational attractive force on μ is equal to the electrostatic repulsive force.

$$G\, M_3\, \mu / R^2 = k\, Q_1\, Q_2 / R^2 = F_0 \quad \text{(2)}$$

R is the radius of the progenitor star.

k is a constant = 9.0×10^9 N m^2 C^{-2}

Causation of the explosion was a mishap: (M_2) was spun off from the central core with velocity V_2. The remaining central core mass (M_1) recoils with velocity V_1. The sudden loss of the (required) gravitational attraction creates a force unbalance for the subsurface nuclear components (μ) by creating an excess Coulomb repulsive force (F_c) which is equal to the lost gravitational attraction. This event, (as seen by astronomers) is the supernova.

$$\text{The force } F_c = G\, \mu\, M_2 / R^2 \quad \text{(3)}$$

The nuclear core is still producing SNFRs (despite the explosion of 1054 C.E). One such group of SNFRs is postulated to revolve around the core in an elliptic orbit, with velocity (v), (assumed to be 1/10 c). (See Fig.10.1)

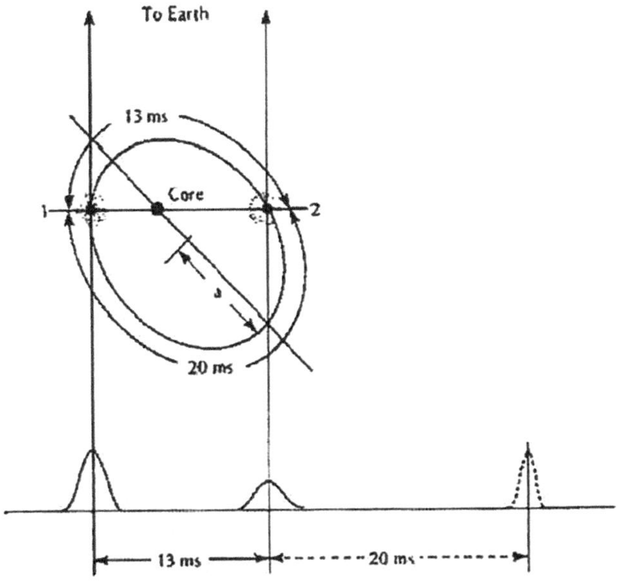

Figure 10.1 Schematic projection of the Crab Nebula pulsar NP 0532. Pulses are presumably emitted when the group of SNFRs is in either sites 1 or 2.

- Let the initial period of revolution of a revolving SNFR about the common center of mass be p_0 (extrapolated from an estimated 900-year pulsar signal slowdown amounting to 0.012 sec).
  $p_0 = 0.0334033$ sec—0.012 sec
  $p_0 = 0.0214$ sec (pulsar period at 1054 C.E.)
- Assume an "effective" orbital radius (a) of an SNFR group revolving around its nuclear core (the pulsar).
  Then, $2\pi a = (0.0214)v$, and $v = (1/10)c$
  Substituting numerical values, $\underline{a = 1.02 \times 10^5 m}$. (Subsequent to the M_2 ejection phase.)
- Assume that pulsar signals originate from orbiting SNFRs with a period of p_0.
  Then, from Kepler's 3rd law, substituting numerical values into its equation:

$$(M_1) = \frac{a^3}{p^2} \bullet \frac{4\pi^2}{G} = 1.37 \times 10^{30} kg \approx 0.68\, M_\odot \quad \text{...... (4)}$$

Conservation of Linear Momentum requires that:

$$M_2\,V_2 = M_1\,V_1 \quad \text{....... (5)}$$

Based on Crab Nebula's Pulsar's recoil velocity of 123 km/s and its mass (M_1), equation (5) can be applied to evaluate M_2, provided we can obtain a value for V_2

$$123 \bullet M_1 = V_2 \bullet M_2 \quad \text{....... (6)}$$

In order to evaluate V_2, a more detailed examination of M_2's ejection process is required. We shall assume that the radius **r** of the spinning nuclear core is slightly less than its orbiting SNFR (about 1.0x 10^5 m).

The standard expression for obtaining the threshold escape velocity (V_{esc}) for an object of mass M is:

$$V_{esc} = (2\,G\,M\,/\,\mathbf{r})^{1/2} \quad \text{....... (7)}$$

Where $M_1 = 1.37 \times 10^{30}$ kg
and r = 1.0x 10^5 m
$G = 6.67 \times 10^{-11}$
Substituting numerical values into equation (7),

$$V_{esc} = 4.27 \times 10^7 \text{ m/s} \quad \text{....... (8)}$$

M_2's velocity must have been in excess (greater than a threshold V_{esc}) initial velocity to escape the pulsar core's gravitational field and achieve a hyperbolic orbit **around** and **below** the nuclear core.

Furthermore, the spin axis of the pulsar would be perpendicular to both its direction of motion and the **E** vector (the direction of motion of the pulsar is in the northwest direction). This also explains the origin of the pulsar's sub-pulses. Since the SNFRs revolve around the nuclear core in a highly elliptical orbit, their polarized electric dipole radiation produces an asymmetric sub-pulse in addition to the main pulse, as shown in Fig 9.1.

Equation (6) will now be used to evaluate the Mass M_2. Substituting numerical values :

$V_2 = 4.27 \times 10^7$ m/s
$M_1 = 1.37 \times 10^{30}$ kg
$V_1 = 123$ km/s provides the mass $\underline{M_2 = 3.94 \times 10^{27} \text{ kg}}$
What was the radius of the original progenitor star?

The acceleration of the total ejected μ system, evaluated for this Supernova event at 0.82×10^{-5} m/s^{-2}) can be applied to Newton's Law of F = ma

Equation (3) becomes:

$$G\,\mu\,M_2\,/\,R^2 = 0.82 \times 10^{-5} \bullet \mu \quad \text{....... (9)}$$

Substituting numerical values into equation (9) (which becomes independent of μ)

$R = 5.66 \times 10^8$ m = $\underline{0.8 \times R_\odot}$

Note that the radius and mass of the progenitor star are comparable to our Sun: 80% and 68% respectively. They thus constitute very reasonable values.

The spun-off mass M_2, subsequent to its ejection, undergoes further evolution via sequential spin-offs of rotating SNFRs, each of which produces a magnetic field. Each SNFR also emits electrons (in addition to positrons, neutrons, gamma rays, and neutrinos). The emission of electrons into the magnetic field is the excitation mechanism which results in the underlying synchrotron radiation.

Based on our present evaluation of the nuclear core's size and mass, its average density is calculated to be: $\underline{3.3 \times 10^{11} \text{ gm/cc}}$: A value consistent as a high density nuclear pulsar. However, it is three orders of magnitude less than a core consisting of pure neutrons, which would be in 10^{14} gm/cc range (according to the Fermi Liquid drop model). In the event that the "progenitor" star would have started its development as a neutron star, it suggests that partial conversion of neutrons into atomic elements must already have occurred in addition to possibly providing a time line for its development.

A Work-Conservation of Energy theorem applied to μ, would be:

$$\text{Kinetic Energy} = 1/2\,\mu\,v^2 = \int \boldsymbol{F} \cdot \boldsymbol{S} \quad \text{....... (10)}$$

where **F** = applied average force and **S** = effective distance over which **F** operates.

Numerical values were not utilized in eqn. 10, since its applicability is questionable in this case. The present size of the nebula is 2×10^{16} m (about two light years). Clearly, the nebulosity has the potential to increase its size.

Overview

The detailed calculations of the Crab Nebula's unique evolutionary development require a summary. The explosive expulsion of a mass M_2 from the original stellar core started the evolutionary process comprising the whole nebulosity. The causation for this process was an internal excess of an electrostatic repulsive force, which in turn resulted in the ejection of the outer layers of the progenitor star. The ejected material included thousands of SNFRs plus a large quantity of hydrogen, helium, and other atomic species. The excitation of the nebulous gas arises from the underlying highly active, nuclear SNFRs, which are in fact nuclear fission reactors, emitting electrons, protons, etc. which produce the excitation.

Meanwhile, the ejected mass M_2 undergoes a hyperbolic orbit below the new nebular structure and its associated gas cloud. M_2 is spinning and ejecting a multitude of SNFRs, which produce their own magnetic fields and at the same time emit electrons, protons, and so on. Thus, this large conglomerate of underlying SNFRs produces the synchrotron radiation that is observed right below the nebulosity. Both sets of SNFR s are essentially moving with an approximately similar distribution of velocities. This then produces the "sandwich" structure, where the upper layer is the emitting gas cloud, and the lower layer is visualized as synchrotron radiation (where both have non-visible SNFRs that produce the excitation mechanism). These two entities are not very close to each other, and with careful analysis can be seen to be slightly separating as the expansion progresses.

Conclusion

The Crab Nebula's numerous enigmas had previously been left to fester for a long time. The above (forensic) analysis, however, demonstrates that order of magnitude calculations can produce consistent and meaningful results when established Newtonian Physics is invoked and effectively utilized.

References

Bejger, M., Haensel, P., 2003, *A&A* **405**, 747–751.

Bietenholz, M.F., Kronberg, P. P., 1991 June 1, Ap. J, **373**:L59-L62.

Burke, B. F. & Franklin, K. L. 1955, J. Geophys. Res., vol. 60, pp. 213–217.

Johnson, F.M., Townes, C.H., 1957, Ap. J, **26**, 466.

Johnson, F.M., 1972, *Mémoires De La Société Royale Des Sciences De Liége*, **3**, 609.

Johnson, F.M., 2015, *Voyage into Astronomy 2nd Ed.*, Lawyers & Judges Publishing Co., Inc., 12, 249.

Johnson, F.M., 2018, *Planetary and Space Science*, **151**, 125–129

Johnson, F.M., 2019, In Preparation.

Mayall, N. U., and Oort, J.H., 1942, *Pub. A.S.P.*, **54**, 95.

Nugent, R., 1998, PASP, **110**, 831.

Oort, J. H., and Walraven, T., 1956, *B.A.N.*, **12**, 285.

Trimple, V., 1968, AJ, **73**, 535.

Trimple, V., 1982, *Rev. Mod. Phy.* **54**, 1183.

Wyckoff, S. and Murray C. A., 1977, Mon. *Mot. R. Soc* ***80***, *717–729.*

Chapter 11

Galaxies

This chapter introduces a new theory which postulates that the motions of stars in a galaxy are under the influence of both gravitational and Coulomb forces (i.e. they exist within a hybrid system). One of the mysteries regarding the non-Keplerian motion of matter in the extremities in the galaxy can now readily be accounted for by a specific type of long-range electrostatic force.

Credit: NASA/JPL-Caltech, An Infrared View of the M81 Galaxy, September 4, 2019

Synopsis

Galactic Structure (Dark Matter Does Not Matter)

Extensive studies of a large number of galactic structures using the well-known Doppler Effect have shown that the outer regions of galaxies do not obey Kepler's laws of planetary motion. Rather than obeying the Inverse Square Law relationship $\left(\frac{1}{r^2}\right)$ the outer regions of a galaxy move according to a $\left(\frac{1}{r}\right)$ relationship with respect to the center of the galaxy. That is, the outer regions of a galaxy revolve like the spokes of a wagon wheel, rather than by the well-known Keplerian relationships. In order to account for this strange anomaly, astronomers have proposed the existence of unknown, unidentified matter which they call "dark matter". They have struggled for decades searching for this elusive, hypothetical, nonexistent matter, resulting in a large number of nonsensical theories.

The Spiral Arms of the Galaxy

The two enigmas associated with the well-measured stellar revolutions in galactic structures:

1. As mentioned above, stellar motions in the outer regions of a galaxy do not obey the standard Kepler (Newtonian Physics) formulation of circular orbital velocity (v) according to the expected relationship $v = \left(\frac{GM}{r}\right)^{\frac{1}{2}}$ but act essentially according to $v = (GM)^{\frac{1}{2}}$, which is independent of the distance r from the galactic center. As a temporary fix, unknown matter (which so far has proven highly illusory) was postulated.
2. The second enigma relates to the diversity of revolving galactic structures, which could consist of a number of spiral arms, each of which is successively further removed from the galactic center.

By just relaxing the stringent but unnecessary requirement that gravity alone reigns exclusively in the universe, a different scenario can be constructed, which lead to a solution of these longstanding dilemmas:

Assume that the central galactic core consists of a dense, electrically-charged core surrounded by one or more similarly-charged masses, which revolve around the core with speed (V) and acceleration (a), toward the center. Then, according to electromagnetic theory, the revolving accelerated charge (Q) produces both a radial field, at a distance r, given by:

$$E_r = \frac{Q}{r^2}$$

And a tangential electric field given by:

$$E_t = \frac{Q \cdot a}{c^2 \cdot r} \cdot sin\theta$$

Where θ is the angle subtended by the radius vector and the direction of the emerging electric field and a is the acceleration produced by the pair of revolving core charges Q. Thus, within the transition shell there is a tangential electric component E_t, which falls off, as r increases, only as the first power of r. This component lies in a plane drawn through the charge and containing the direction of its acceleration (a). As time goes on and $r \rightarrow \infty$, E_r will become entirely negligible in comparison with E_t (Richtmyer and Kennard 1947, Pg. 63).

Here finally is a physical effect that can be applied to the enigmatic motion of galaxies and all their varied structures. The universe operates under a ***hybrid*** system, composed of both gravitational and electromagnetic force fields! It is important to realize that, should this hybrid system exist in our universe, then it must follow that all galactic objects are also potentially electrically charged (not just galactic cores).

As such, the tangential electric field component of the hybrid system will produce the observed outer spiral ring structure as it radiates out from the central core to interact with the galaxy's stars and electrically-charged gas clouds.

Assuming this "hybrid" effect exists within all galaxies, the tangential electric field components radiating out from the central dynamic core configuration will produce the observed outer spiral ring structures as it interacts with each of the outlying stars as well as electrically-charged gas clouds. As the core of the galaxy evolves, it ejects mass that forms proto-stars. Conservation of angular momentum insures that each of the orbiting masses lands in a different orbital radius. The eccentricity of the resulting orbit depends on the manner in which its proto-star ejection process evolved. (One other thought: our solar system's major planets, on a smaller scale, developed in a similar fashion.)

One should now be in a position to develop the large variety of galactic structures based on the evolutionary development of a galaxy's massive electric core and the unique development of its substructures. A survey of some of the photographs of galaxies shows a spherically symmetric central distribution resulting from the $\frac{1}{r^2}$ gravitational force field relationship and the outer flat-type spiral structures, which are a result of the $\frac{1}{r}$ electric force field relationship, which lies in a plane, providing the flat, disc-type spiral configurations, as predicted by the theory outlined here (Figure 11.1).

Galaxy M74

NGC 6946: The 'Fireworks Galaxy'

M51, also known as NGC 5194
The Whirlpool Galaxy

UGC 6093

Figure 11.1 *Credit: NASA Photographs*

For special emphasis, this important theoretical development scenario is mentioned here then reintroduced in chapter 12.

Corollary

1. One may therefore postulate that not only galaxies, but all stars operate on this hybrid system.
2. Based on this new theoretical model of a hybrid gravity-electromagnetic field development of a galaxy, it clearly does not allow for the existence of a theoretical "Black Hole" situated at a galaxy's center (which incidentally would be incapable of producing galactic structures, based on observations).
3. In fact, this hybrid model sounds the death knell for all so-called "Black Holes" as well as illusionary postulated Accretion Discs that are alleged to swallow up dozens of solar masses per year!

References

1. Hutchings, J.B., Gower, A.C., Price, R. The Astronomical Journal **92** (1987).

2. Kanekar, N., Prochaska, J.X., Ellison, S.L., Chengalur, J.N. *Monthly Notices of the Royal Astronomical Society* **396**, 385 (2009).

Chapter 12

A Modern Overview of Stellar & Galactic Evolution Scenarios

Synopsis

The two principles listed below perform a critical function in explaining two enigmas in astronomy.

The (Johnson) Similarity Principle

All galactic nuclei, all stars, and all planets contain within their structure spinning cores and include at least one asymmetric charge located at each of their core's periphery. Alternatively, a revolving binary pair of cores, with one of them carrying an excess electric charge, could also constitute possible internal structures (which could either be a galactic center or a stellar core configuration). All internal energy sources are by means of nuclear fission (Planetary and Space Science 151, 2018, 125–129). An enormous number of fission reactors are assembled beneath outer surfaces of galactic nuclei and also below stellar outer surfaces. Equilibrium is maintained by a combination of electric coulomb repulsion and gravity core attraction of these complex nuclear assemblies.

Corollary

Observed magnetic field reversals by the Sun and the Earth confirm the central core's magnetic field and their numerous flip-overs.

An (Old) Electromagnetic Phenomenon (revived as a solution in Astrophysics)

Consider the radiation emitted from an accelerated point charge (See p. 60–63 Introduction to Modern Physics by Richtmyer and Kennard, 4th Edition, 1947; also p. 227 Electricity and Magnetism by Abraham and Becker, 2nd Edition, 1930).

Let q be a point charge with acceleration (a) making an angle (θ) in the direction of its intended target (P) at a distance (r). Then there will be two components of the electric field at P: A radial component

$$E_r = \frac{q}{r^2}$$

and a tangential electric field component E_t Where

$$E_t = \frac{qasin\theta}{c^2r}.$$

"Note that the tangential electric field component falls off, as r increases only as the first power of r. This component lies in the plane drawn through the charge and containing the direction of its acceleration. As time goes on and r→∞, E_r will become entirely negligible in comparison with E_t. Thus, ultimately, we have a shell or pulse of field, traveling outward with the speed of light, in which the electric vector is transverse, just as in plane waves."

The total energy radiated per second, over a sphere, is given by this expression:

$$\Omega = \frac{2q^2a^2}{3c^3}$$

"According to classical theory, an accelerated charge must necessarily radiate energy." (See page 66 of Richtmyer and Kennard).

Introduction

Astronomy has clearly made enormous strides during the last century, primarily by advances in the range of energy detection and by launching a host of multifunctional precision satellites. (The enormity of this enterprise has simultaneously generated an increase in the number of practitioners.) In view of enhanced data acquisition-processing and advances in technology, a number of puzzling anomalies have come to the surface. Among these anomalies was a particularly vexing one, known as the "Pioneer Anomaly". An older enigma concerns stellar velocities in outlying galaxies.

The two enunciated principles will be essential in providing solutions to these enigmas.

We shall first reexamine the seemingly small but potentially important Pioneer anomaly:

Pioneer 10 and Pioneer 11 spacecraft were launched in 1972 and1973 respectively, intended to exit the solar system in roughly opposite directions. At heliocentric distances in the 20–70 AU range, the presence of an anomalous average sunward acceleration of $a_p = (8.74 \pm 1.33) \times 10^{-10}$ m/s^2 was measured for both vehicles. This anomaly festered unexplained for over a decade. A thermal recoil force explanation was eventually put forward in order to perhaps remove this anomaly from further study, since its continued presence was a possible embarrassment. However, there are a number of other puzzling, unexplained anomalies associated with precision satellite tracking (mostly near planetary objects). These unresolved enigmas include five satellite-Earth flybys involving gravitational assist events in addition to the galactic stellar flat rotation curves. From a scientist's point of view, precision satellite tracking should be appreciated as sensitive test probes rather than have its potential for new physical phenomena being overlooked, according to Paul ten Boom. He also argued strongly to not disregard the positive opinion of the team leader John P. Anderson, who was extremely skeptical of the thermal recoil force hypothesis. In keeping with this spirit, there is a novel alternative hypothesis that had so far not been considered, since the topic requires knowledge of this author's theories, his novel principles and our Sun's unique development scenario (See textbook *Voyage into Astronomy* 2015 and *Planetary and Space Sciences* 2018).

The reported **Pioneer spacecraft anomaly** (a_p) is constant over an approximate range of 25–70 AU for Pioneer 10 and for a smaller range for Pioneer 11. Power to each spacecraft was provided by four SNAP-19 (Space Nuclear Auxiliary Power) radioisotope thermoelectric generators. The energy arises from radioactive Pu238 whose alpha emission interacts with thermoelectric generators. The four radioactive thermal generators (RTGs) delivered 160 watts of electrical power at launch. Each emitted alpha particle carries a positive electric charge, which accumulates somewhere in the satellite. The total size of this charge was estimated by (Anderson *et al.* 1998, 2002, 2008) at a value of q=0.1 μC, which in turn produces an electric field. The mass of the Pioneer spacecraft was 241 kg (with 50% fuel left), when the canonical sunward acceleration (a_p) was measured at 8.74×10^{-10} m/s^2.

Hence, using Newton's (F=ma) force law: F= $241 \times 8.74 \times 10^{-10}$ N towards the Sun.

$$F = 2.1 \times 10^{-7}\ \text{N}$$

The accumulated electric charge (q) in the power generator system was equal to 10^{-7} C. Hence, the corresponding electric field, which is defined as force per unit charge, becomes 2.1 N$\times$C^{-1}. The presence of this electric charge plays a critical role in the final interpretation of this enigma.

Two characteristics of the Pioneer Anomaly had previously remained unexplained.

1. Only low-mass bodies seem to be effected ($<10^{14}$ kg).
2. Beyond a distance of 20 AU, the main value of the anomaly is roughly constant. (Both effects are now accounted for by the Johnson hypothesis).
3. There is an "annual" variation structure superimposed upon the mean constant value a_p, which is greater than the measurement error. Its annual period is about 355 days, i.e. 10 (days) degrees short of a complete 365.25 day Earth orbit.

An (oversimplified) thermal, non-isotropic heat radiation effect during the satellite's flying time was a proposed solution. This asymmetric heat radiation explanation, however, failed to account for the small annual variations superimposed upon $a_{p.}$ The observed effect implies that the Earth's revolution about the Sun had contributed to this superimposed variation and furthermore, its ten-degree offset is readily explicable by the movement of the Sun during a one year period travelling at a speed of twenty kilometers per second. In fact, the angle subtended from a distant Pioneer satellite to the center of the Sun during a one year period will produce a triangle whose base is the travel distance by the Sun during the course of a year which amounts to 6.3×10^{11} m. Taking the average distance of the Satellite from the Sun at 25 AU, it would correspond to 37.5×10^{11} m. The apex of this triangle would be exactly ten degrees (0.17 radians), thus confirming unambiguously that the effect originates within the Sun and that the Earth, by virtue of its revolution about the Sun, contributes a very small addition to this anomaly.

This analysis implies that we have the superposition of two separate effects (the Corollary of the Similarity Principle): A specific charge rotation within the Sun's core causing acceleration towards its center and a similar but much smaller effect originating with the revolution of the Earth's central charge about the Sun. Both of the accelerated charges produce a tangential electric field, interacting with the accumulated charges on the pioneer satellites, and thereby introduce the small observed acceleration of a_p towards the center of the Sun.

The constancy of the acceleration ($\mathbf{a}_p$) in the range of 20 AU to about 70 AU is observed despite the $1/r$ effect of the tangential electric field. However, since the accumulated electric charge increases at a rate proportional to its distance (r) traveled, the two effects cancel each other out, and $\mathbf{a}_p$ remains constant.

Evolution of Galaxies

Dumbbell Hybrid Systems

A typical galaxy starts its lifecycle as a massive, rapidly-spinning dense nuclear core. Due to its instability in its extremely rapidly-spinning state, it will split into two spinning electrically-charged objects, revolving around their common center of mass to form a "dumbbell" system, whose configuration we will now discuss. Splitting phenomena can also occur in stars: High velocity splitting of an internal revolving dumbbell results in 2 separate stars (a binary star system). While a minor splitting produces an object characterized as a variable star (a Cepheid), whose oscillations (P) are extremely precise, clearly indicative of an internal dumbbell configuration within the star. This phenomenon is theoretically confirmed. Cepheid's measured periods (p) are inversely proportional to the square root of their average density (ρ). Kepler's Third Law:

$$\left(\frac{P^2}{a^3} = \frac{4\pi^2}{G\left(m_1 + m_2\right)}\right).$$

By assuming that the masses of the internal dumbbell exceed the mass of its surrounding stellar gas environment, one can readily deduce from this equation that the period (p) is proportional to $\frac{1}{\sqrt{\rho}}$. Where ρ is the density and $\boldsymbol{a}$ is the mean distance of its semi-major axis.

Principle No. 2 for the specific dumbbell case. (This unique effect was known for about a century, here it is restated more explicitly).

The transverse electric field component term (E_t) for all dumbbell cases is:

$$E_t = \frac{qa\sin\theta}{rc^2}$$

The acceleration (a) corresponds to the internal revolution of the dumbbell system about its center of mass, (q) is its charge, and (r) is the distance from the dumbbell system to the point to be measured. θ is the angle between the direction of the accelerated charge and the direction towards a given star.

Consider now galaxies using Principle No. 2, assuming that each of the dumbbell masses carries a positive electric charge. There is the gravitational attractive force acting between dumbbells and its outlying stars diminished by their inverse square relationships, according to Newton's Laws. However, in addition, there *now* exists an additional pair of electric fields emanating from the spinning electrically charged dumbbell. One of them is a radial electric field, which drops off as per Inverse Square of the distance from the core to an outlying star. The other is an additional transverse electric field component, (E_t) which falls off as $1/r$ as its distance from a star increases. Thus, this transverse electric field component (E_t) will reach out much further into distance space, while the radial component becomes entirely negligible. Moreover, this transverse electric field (E_t) lies in the plane of and parallel to the dumbbell's revolution. **This theory will account for the observed flat structure of the outlying arms of a galaxy**.

Massive spinning cores within each galaxy begin their development, via the dumbbell mode. Its stars are launched by means of the *hybrid system*, which ensures flat rotation of galactic structures. Once all stars are launched, however, feeble tangential electric field components could still remain in operation, acting with very weak forces on outlying objects. In fact, for our galaxy, they may have completely vanished. Stars situated in the spiral arms of other galaxies, however, could, but need not necessarily depart a galaxy during its unwinding process.

The tangential electric field component, which operated in our galaxy at its initial birth development phase, is no longer operating at this time, since it would have adversely affected the distant Pioneer satellites 10 and 11.

Summery

Our galaxy originated from a very dense, heavy nuclear core. It developed a binary system which contained a substantial electric charge during its evolutionally and stellar ejection processes. The ejected proto-stars carried also an electric charge and were interacting with the tangential electric field component generated by the galactic cores binary revolution. This not only produced the necessary $\frac{1}{r}$ relationship but also ensured that the tangential electric field, which is in the plane of the predominant binary rotator, produced a relatively flat galactic structure.

A Classical Newtonian Physics Analysis of our Galaxy

The most recent astronomical data have established the mass of the center of our galaxy as 8.2×10^{36} kg, at a distance of 2.46×10^{20} m from our Sun [Ghez et al.2008]. The galaxy center is also a very powerful radio emitter, known as SgrA*.

We will first calculate the necessary force for keeping our solar system in circular orbit. Given the speed of the Sun as 20 km per second and using the following numerical parameters.

v = 20 km/s (space velocity of the Sun)

$R_0 = 2.46 \times 10^{20}$ m (8.0 kpc) (distance of Sun to galaxy's center)

$M_0 = 2 \times 10^{30}$ kg (mass of the Sun)

$M_b = 8.2 \times 10^{36}$ kg (mass of the galactic center)

$G = 6.672 \times 10^{-11}$ $m^3 kg^{-1} sec^{-2}$

The necessary force needed to keep the Sun in its orbit is $\frac{M_0 v^2}{R_0} = 3.25 \times 10^{18}$ N.

If only the core were involved, it would produce an insufficient force to keep the Sun in orbit which would be $G\frac{M_0 M_b}{R_o^2} = 1.81 \times 10^{16}$ N.

However, all the stars in our galaxy that are *within* the Sun's circular orbit contribute their gravitational effects as well. Thus,

$$M_{total} = \frac{v^2 R_0}{G} = 1.47 \times 10^{39} \text{kg} = 2.3 \times 10^8 M_0$$

According to this simple calculation, there are about 230 million stars (assuming that their mass is comparable to that of our Sun) within our Sun's orbit. This would include the equivalent mass of 4.1 million Suns associated within our galactic center (mischaracterized as a Black Hole).

Based on the above analysis, Black Holes cannot develop: It is based on a theory which assumes that all matter in the universe is charge neutral. This is an ad-hoc, unsubstantiate assumption. Internal electric fields or large numbers of complex structures (e.g. SNFRs) within a massive object, prevent gravitational collapse scenarios.

References

Abuter, R., *et al.*, 2017, A&A **602**, A94 (GRAVITY Collaboration)

Abuter, R., *et al.*, 2018, A&A **615**, L15 (GRAVITY Collaboration)

Allen's Astrophysical Quantities 4th Edition, 1999, A.N. Cox Ed.

Anderson, J.D., *et al.*, 1998, Physical Review Letters **81**, 14

Anderson, J.D., *et al.*, 2002, Physical Review D **65**, 082004

Anderson, J.D., *et al.*, 2008, Physical Review Letters **100**, 091102

Ghez, A.M., *et al.*, 2005, Ap.J **620**, 744

Ghez, A.M., *et al.*, 2008, Ap.J **689**, 1044

Hill, F., 1999, Allen's Astrophysical Quantities 4th Edition, Solar Oscillations, 342

Katz, J.I., 1999, Physical Review Letters **83**, 1892

Kerr, R.A., 2011, Science **333**, 1208

Livingston, W.C., 1999, Allen's Astrophysical Quantities 4th Edition, Basic Data, 340

Muller, R., 1999, Allen's Astrophysical Quantities 4th Edition, Granulation of the Sun's Surface, 364

Turyshev, S.G.et al. 2006, International Journal of Modern Physics D **15**,1–55

Turyshev,S.G and Toth, T., Living Reviews in Relativity **13**, 1–175

Chapter 13

SgrA*: The Center of Our Galaxy

The search for and study of our galactic center with all its possible scientific ramifications has a long history. Large amounts of dust in the intervening space between us and the galactic center had prevented visual telescopic observations. Yet, the center also hosts a very bright, distinct radio source known as SgrA*. In the years between 2000 and 2017, very long baseline mm wave interferometry was able to locate and measure the actual size of this radio source, having a radius of 2.1×10^{10} m (30 times bigger than our Sun). Meanwhile, on the summit of Mauna Kea, UCLA astronomers at the Keck 10m telescope, (equipped with the most advanced infrared (2-micron infrared K band) detection equipment) were able to penetrate the obscured dust and facilitate its detailed study. They located a collection of stars (45 so far) moving in an assortment of diverse (clockwise and anti-clockwise) orbits about the center, SgrA* (2.5×10^{17} km from us). Tracking the orbits of 17 such stars for about a dozen years, the mass of SgrA* was determined to be 4×10^6 greater than that of our Sun. An object of this immense mass was immediately assumed to be a Supermassive Black Hole by UCLA astronomers. However, by prematurely assuming that the galactic center harbors a Black Hole, two enigmas became immediately apparent and inexplicable: How can stars develop from a collapsing dust cloud in the immediate vicinity of an object that theoretically would consume it? SgrA* is also the origin of intense X-rays, variable infrared radiation, and polarized electromagnetic flares moving at extremely high velocities. How can these diverse complex signals be generated when one postulates an object that would theoretically consume rather than violently emit these signals? A simpler scientific explanation (based on the most recent discoveries) does exist, as will be shown below.

Of all the stars circulating the massive central object (SgrA*), one particular star (SO2) caught astronomers' attention in view of its elongated eccentric orbit and getting as close as 120 AU (it's pericenter) during its 16-year orbit around SgrA*. A major scientific effort was mobilized, which included the following: The world's largest telescopes (10m Keck, New Technology Telescope, and the Very Large Telescope), the most advanced sophisticated technology and the combined efforts of three separate groups (Max Planck (MPE), University of Cologne (comprising the Gravity Collaboration) and UCLA. All of it focused on Star SO2. Why this star? It was relatively the brightest and the closest star to SgrA*. The 26-year orbital motion study of SO2 utilized high resolution near-infrared (NIR) speckle and adaptive optics, assisted imaging and spectroscopy for velocity measurements. It resulted in establishing SO2's precise orbit and its speed, particularly at its closest approach to the galactic center. While SO2's average orbital velocity is 1898 km/s, at the pericenter (its closest approach to SgrA*), its space velocity reached 7650 km/s. This peak velocity, however, was about 230 km/s too large, according to standard Newtonian orbit analysis (Abuter et al. 2018).

A scenario similar to that of our planet Mercury and our Sun, a relatively young B-type star (SO2) is testing the gravitational force in its closest approach to SgrA*. It discovered that Newtonian physics had missed the gravitational effects of another, far larger, hidden subsurface Q, in roughly the same manner as it was shown for the Sun. Their respective energies are derived from similar nuclear fission processes. Common features included dense, spinning, electrically charged nuclear cores and internal plasmas. Both systems produce periodic spinoffs from their spinning nuclear cores, which spiral towards and settle in at their respective outer sub-surfaces. Their orbital speed is also insufficient for gravitational equilibrium; hence, electrostatic repulsion forces arising from their cores are the required forces to maintain quasi-static equilibrium. It is this gravitationally balanced repulsive electrostatic force which is missed in Newtonian kinematic analysis.

Both types of objects, irrespective of their size and mass, display large quantities of variable electromagnetic emissions, which include IR, x-rays, and the impressive gigantic flares (the Similarity Principle).

The Gravitational Collaboration (Abuter, et al. 2018) made an additional discovery: they utilized four 8 m telescopes (La Silla Paranal Observatory, Chile) to measure the orbital velocity of moving flares at or near outer regions of the galactic centroid. This discovery provided an estimate for the centroid's surface rotational velocity. Similar mechanisms are responsible for the flare's origin.

Based on the Sun's Luminosity of 3.84×10^{23} kW and its active mass Q being 3×10^{23}kg, its output efficiency is 1.3 kW per kg. Since the computed, active mass of Sagittarius's subsurface Q is 2.0×10^{36} kg (Appendix A), its projected Luminosity is 2.6×10^{36} kW. Another method for obtaining the same result is to invoke the "Johnson Similarity Principle" (the Luminosity of an active fission object is proportional to the square of its mass):

Assume that the "average" mass (M_{sgr}) of SgrA* is 5.2×10^{36} kg (instead of 8.2×10^{36}kg), comparing it to known values for the Sun, one obtains the following equation, which compares the luminosity ratio of these objects:

$$L_{sag}/L_{sun} = [M_{sgr}/M_{sun}]^2$$

Inserting numerical values, Lsag = <u>2.6×10^{39} watts</u>.

Both the Sun and SgrA* are assumed to radiate as Black bodies with power **P** (per unit area) using the standard Stefan-Boltzmann relationship: $\mathbf{P=\sigma T^4}$

Where $\sigma=5.67\times10^{-8}$ W per m^2 or in cgs units: $\sigma=5.669\times10^{-5}$ ergs cm^{-2} sec^{-1}

The radius of SgrA* is 30 times larger than the Sun's. Its surface area is therefore 900 greater; (it is 5.54×10^{25} cm^2). Inserting numerical values into $P=\sigma T^4$ and solving for T. By taking the fourth root of T^4, <u>$\mathbf{T=1.7\times10^6}$ **K**</u>

Thus, the surface temperature of SgrA* is similar to the Sun's corona temperature. The major difference being that the density of SgrA*'s corona is very much greater and in fact establishes its surface temperature.

Wien's Law, λ T=0.29 [cm K] will determine the wavelength at which the power is a maximum. Inserting the derived temperature of 1.7×10^6 K, the maximum is at 17×10^{-8} cm (17 Å).

The Stefan-Boltzmann relationship was invoked in order to obtain the spectral energy distribution of SgrA* for the frequency range of 10^{10} to 10^{18} Hz. These calculations were done for the temperature of <u>$\mathbf{1.7\times10^6}$ **K**</u> **and** for 3×10^5 K (for comparison).

The graphical representation (Figure 1) showed that the power spectrum is linearly related to the frequency (viz. $1/\lambda$) for both temperatures, up to their maximum values. Beyond their maximum, the luminosity output power drops steeply.

It is now possible to compare the observed and actual spectral energy distributions: Fig1 and Fig4, (ref. Gravity Collaboration,2020). Note the enormous amount of scattering, relative to the actual spectral energy output of SgrA* (Fig1).

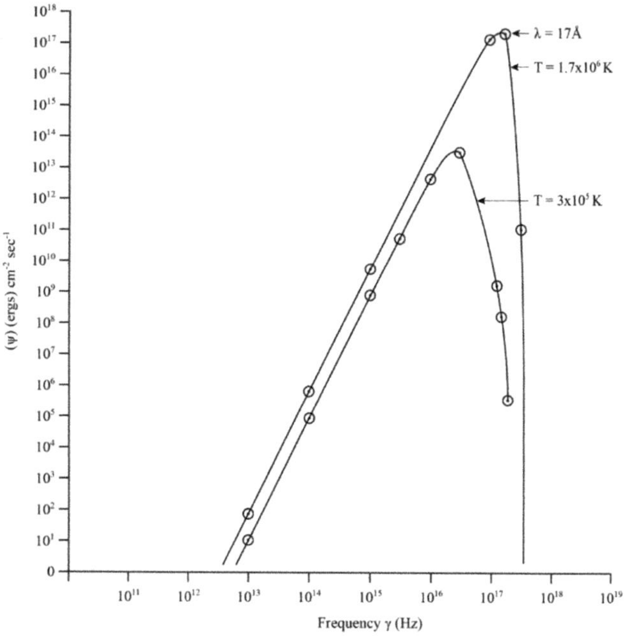

Fig.1 *The Stefan-Boltzmann flux distribution of Sgr A** ***The power output of SgrA* calculated according to the Stefan-Boltzmann relationship for the temperature of 1.7×10^6K*** *(and for 3×10^5K, for comparison)*

The flux distribution of Sgr A* according to the GRAVITY Collaboration R.Abuter et al., A&A 638, A2, (2020)

The Spectral Energy Distribution from Fig 4 below, is compared with the actual power output shown in Fig1

Take the complete spectral energy distribution from Figure 4, and compare it with the Stefan-Boltzmann plot in figure 1, which provides the actual energy source output of SgrA*, relative to the baseline. Interstellar scattering reduced the observed NIR measurements, by a factor of 10^7. The measured x-rays at an average frequency of 10^{17} Hz where reduced by a factor of 10^{20} relative to the base line at 10^{12} Hz.

The average flare intensity at 10^{16} Hz would be a factor of 100,000,000 more intense than the NIR reading at 10^{14} Hz.

The enormous energy output of SgrA*, extending from 10^{12} to 10^{17} Hz, is only explicable by means of a colossal multibillion fission reactor energy complex. Its presence is incompatible with an assumed hypothetical Black Hole.

Conclusions

The **radius** of the mass centroid SgrA* is $\mathbf{2.1\times10^{10}}$ **m**, measured by means of very long baseline mm interferometry (Falcke et al. 2000, Doelman et al. 2008, and Johsnon et al. 2017). Its mass was established (Ghez at al. 2005 and 2008) at $\mathbf{8.2\times10^{36}}$ **kg**. Thus, its mean density is <u>**211 gm/cc.**</u>

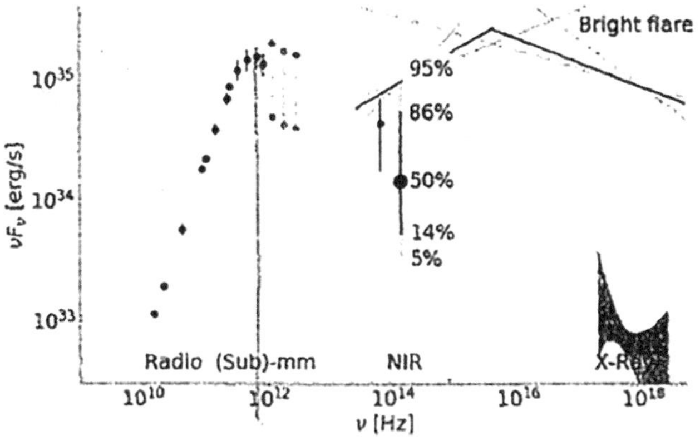

Fig. 4. *SEO of Sgr A*: the radio and sub-mm data are from Falcke ctaT' (l 998), Bower et al. (20 l 5, 2019), Brinkerink et al. (2015), Liu et al. (2016), The far infra-red data is from Stone et aL (2016) and von Fellenberg et aL (20 l 8). The NIR M-band data is the median flux density inferred from the lognormal model of Witzel et al. (2018). The NIR K-band data is the GRAVITY flux density: the thick point is the median flux density, and further flux density percentiles are annotated. Also shown are the NIR and X-ray flux density spectrum of a brig]:it simultaneous flarn observed by Ponti et al. (20l 7), the quiescent X-ray flux density is determined from Baganoff et al. (2003).*

An average density of 211 gm/cc does not support conditions for a hypothetical Black Hole development.

Hybrid (electrostatic/gravity/nuclear) energy mechanisms within any massive object ($M > 100\ M_\odot$) prevents theoretical Black Hole construction (*Gravity,* Hartle 2003).

Thus, according to the Similarity Principle, our galactic center and all other galactic centers do not ever evolve into Black Holes.

A tight-fitting mass-luminosity relationship (Chapt.1) demonstrated that all planets and stars operate by a common energy mechanism (nuclear fission). The Similarity Principle now includes the galactic centers in this list.

1) Newtonian Physics vs. General Relativity (G.R.)

In a Newtonian motion analysis, kinetic forces are equated to the resulting gravitational attraction. Suppose that a small part of the total mass is separated and balanced by an electrostatic attractive force. This would then be neglected in this type of force analysis. Thus, here is a rare example of an exception to the Equivalence Principle. G.R., on the other hand, uses an entirely different approach in its solving procedure: it is energy-based (rather than force-based) since it involves an energy/momentum tensor in its basic complex equations. Hence, it has the ability to seek out the very effect that Newtonian kinematics has missed. The Sun and the galactic center provide two robust examples which illustrate these phenomena in detail: The Sun with the planet Mercury, SgrA* with an orbiting star (SO2). Despite their large differences in magnitude, both phenomena are explicable by the same physical principle. The mathematical details are shown in Appendix A. The attached table summarizes all the relevant data for both the Sun and SgrA*.

There are about 5,000,000 subsurface, electrically charged spinning nuclear fission components on the Sun, having a total mass of 3.3×10^{23} kg. On a far grander scale, the galactic center has a subsurface which contains billions of active nuclear components comprising a total mass of approximately 2×10^{36} ($10^6\ M_\odot$).

Based on Johnson's Similarity Principle, the total luminosity of SgrA* would be in the 10^{36} kW range.

The multiplicity of phenomena we observe on the Sun are replicated but with far greater intensity on the surface of the galactic center. Thus, in view of its enormously large surface electric field, its display of x-rays variable IR and flares should be readily accounted for by SgrA*.

The surface electric fields on both the Sun and SgrA* are a consequence of equilibrium requirements for balancing the electric charge repulsive forces against the gravitational attraction of the outer-subsurface masses. These requirements are present on both the Sun and the galactic center, SgrA*.

On the subsurface area of SgrA* (6.16×10^{18} sq. m) reside billions of nuclear fission components (of unknown dimensions). They would also harbor spinning electric charges. The core of SgrA* (with comparable or higher density) would also carry a large electric charge so that its repulsive force is exactly equal to the gravitational attraction of the core and the subsurface mass (M_Q). The equation describing this condition is as follows:

$$G\ (\text{Mass of core}) \times (\text{mass of } M_Q) = k(Q_1 \times Q_2)$$

Where the total charge on the core is Q_1 and where Q_2 is the total electric charge on M_Q. It is understood that Q_1 and Q_2 are of equal polarity. The inverse square law, of course, applies to both gravity and Coulomb repulsive forces. The electric field strength on the galactic centroid's surface could be a million times greater than on the Sun's surface.

Appendix A

The SO2 orbit about SgrA*

The non-Newtonian red shift explicable by General Relativity and its novel interpretation.

This constitutes the second example of a deficiency in a Newtonian-Kepler motion analysis. It is based on a highly unusual (rare) failure of the Equivalence Principle: A specific, subsurface, complex mass (Q) is in equilibrium by virtue of two (equal in magnitude) opposing forces: an electrostatic repulsion, and a gravitational attractive force, thereby initiating the so-called missing mass Q phenomenon. The mathematical analyses are similar for the Sun's and the galactic center's subsurface structures. They differ merely in their widely different masses and their proportionally different outcomes.

The G.R. anomaly is a 230 km/sec excess speed of star SO2 in its closest approach to the galactic center (SgrA*). A variable speed increase (being a gravitational effect) would be present during all of SO2's orbit. However, its magnitude, being distant dependent, would be most readily noticed during its closest approach to the center.

All the parameters needed for an exact calculation (which includes those for the Sun as well) are displayed in the attached table. Whereas the mass of Mercury was well known, SO2's mass (m) is not well established. Based on its stellar B2.5 classification, we estimate m to be in the 10 $M_\odot$ range ± 30%. Thus, our final result would also reflect this uncertainty.

The average orbital velocity

$$v = \left(\frac{M_B G}{a}\right)^{1/2} = 1898 \text{ km/s} \quad \text{.................................} (1)$$

Where M_B is the mass of SgrA* = 8.2×10^{36} kg

a = semi-major axis of star SO2 = 1.53×10^{14}m

Let T be a Newtonian equivalent "extra" force responsible for the 230 km/s anomaly.

Then, according to an equation derived previously, the increased velocity

$$(\Delta v) = \frac{Ta}{2vm} \quad \text{...} (2)$$

Where m = 2×10^{31} kg (mass of SO2)

$\Delta v = 2.30 \times 10^5$ m/s

Substituting numerical values into equation (2)

$T = 1.14 \times 10^{29}$ N

Let the additional subsurface mass be M_Q

Then, the extra gravitational effective force T (which causes the observed 230 km/s increase) will be used to evaluate the "hidden" mass Q by Newton's law of gravitation:

$$T = \frac{GM_Q \cdot m}{a^2} \quad \text{...} (3)$$

$$M_Q = \frac{Ta^2}{G \cdot m} \quad \text{...} (4)$$

Substituting numerical values into equation (4)

$M_Q = 2.00 \times 10^{36}$ kg

$\underline{M_Q = 10^6\ M_\odot}$

Table

	Sun	Mercury	SO2 (B2.5 Star)	Galactic Center
Total Mass	1.989×10^{30} kg	3.30×10^{23} kg	2×10^{31} kg (10 $M_\odot$)	8.2×10^{36} kg
Surface Radius	6.955×10^8m			2.1×10^{10} m (0.14 AU)
Average Density <p>	1.409 gm/cc	5.43 gm/cc		212 gm/cc
Inner Core Radius*	2.179×10^8			(6.61×10^9 m)**
Core Mass *	5.637×10^{29} kg			(2.33×10^{36} kg)**
Core Density <p> *	>13 gm/cc			1927 gm/cc
Semi-Major axis (a)		57.9×10^9 m (0.337 AU)	1.53×10^{14} m (1023 AU)	
Pericenter R			1.79×10^{13} m (120 AU)	
Orbital <v>		47.87 km/s	1898 km/s	

Table

	Sun	Mercury	SO2 (B2.5 Star)	Galactic Center
Pericenter v			7,675 km/s	
Δv at Pericenter		4.4 mm/s (Continuous)	230 km/s	
Mass at Surface (Q)	3.3×10^{23} kg			2.0×10^{36} kg ($10^6 M_\odot$)
Surface Rotl. v	2 km/s			4.88×10^{7} m/s
Equilibrium v	437 km/s			1.6×10^{8} m/s
Eccentricity		0.205	0.8847	
Galactic Center R				2.5×10^{20} m (8.122 kpc)
(SO2) Period			16.05 yrs.	Spin rate: 45 min.
$G=6.672\times10^{-11}$ m^{3}kg^{-1}s^{-2}				
1 AU $=1.4959\times10^{11}$ m				
1 pc $=3.085\times10^{16}$ m				

* Similarity Principle

** Preliminary Estimates

References

Abuter, R., *et al.*, 2017, A&A **602**, A94 (GRAVITY Collaboration)

Abuter, R., *et al.*, 2018, A&A **615**, L15 (GRAVITY Collaboration)

Abuter,R et al., A&A 638, A2, (2020)

Allen's Astrophysical Quantities 4th Edition, 1999, A.N. Cox Ed.

Doeleman et al. S.S, 2008, Nature **455**, 78

Falcke, H., et al., 2000, ApJ. **528**, L13

Ghez, A.M., *et al.*, 2005, Ap.J **620**, 744

Ghez, A.M., *et al.*, 2008, Ap.J **689**, 1044

Hill, F., 1999, Allen's Astrophysical Quantities 4th Edition, Solar Oscillations, 342

Johnson, M.D et al., 2017, AptJ. **850**, 172

Katz, J.I., 1999, Physical Review Letters **83**, 1892

Kerr, R.A., 2011, Science **333**, 1208

Livingston, W.C., 1999, Allen's Astrophysical Quantities 4th Edition, Basic Data, 340

Muller, R., 1999, Allen's Astrophysical Quantities 4th Edition, Granulation of the Sun's Surface, 364

Ponti, G., *et al.*, 2017, MNRAS **468**, 2447

Chapter 14

Venus & Global Warming

Synopsis
A Discussion of Water
Solubility of Water
Venus
References

A Discussion of Water

Could live have ever started on Mars? Although the great emphasis now is on exploring Mars, the planet Venus seems to be completely neglected. We know that one of the most essential conditions for life as we know it is the presence of water.

Keep in mind, however, that a great deal of life at the lower evolutionary scale can exist with very little water, e.g., type Kaoline termites, to the consternation of southern California homeowners. These termites survive from water extracted from the moisture in the air. Nature has shown us on earth many unusual, but successful, adaptations of life to very dry and harsh conditions: deserts support a great diversity of living systems, including a species of rats (mammals) which survive without drinking a single drop of water. These rats obtain water from plants or internally, through body carbohydrate processing. This knowledge leads one to have an open mind about previous life on Mars. We may have an answer soon, but if the answer is negative, remember this: what are the chances of finding another human being if you were suddenly placed anywhere among the vast unpopulated regions of the world? (We are now 7 billion people, with at least 20% of the solid earth's surface uninhabited and inhospitable.)

Let's now examine some of the other properties of water. How much of it is there on earth? About 97% of all water is tied up in the oceans. The bulk of the remainder is in the form of ice. Only I part of water out of 2,000 parts of ocean water is in the form of available fresh water, and only about 1/30 of all the fresh water is in the vapor state (including clouds). These numbers show how little fresh water is actually in circulation to sustain life. The ocean is an enormous, nondrinkable reservoir. The Sun's energy performs the vaporization, or distillation, process, thereby producing the pure water which ends up as clouds and eventually produces our needed rain.

Water requirements:

1 person requires 2,000 gallons/day or 7×10^6 cc per day. (Not all is for drinking! Water is mainly needed for growing food.)

Population of 7.8 billion requires:

$$7 \times 10^9 \times 7.8 \times 10^6 = 5.5 \times 10^{16} \text{ cubic cm of water per day.}$$

Total amount of water from rain and run-off $= 3.6 \times 10^{17}$ cc

Thus, world population (unevenly distributed) water requirements seem to approach 15% of its total circulating supply!!

Solubility of Water

The solubility of water is another property indispensable for life. Table A gives the major dissolved constituents in sea water. The ability of water to dissolve a large number of chemicals allows for erosion of the earth's surface, bringing essential chemicals and minerals with the flow of rivers into lakes and oceans, thus supporting life. Man's interference in these natural processes has immediate deleterious consequences, as for instance the construction of the Aswan Dam. This dam now prevents the previous flooding and nutrient-bearing lower Nile from fertilizing and "feeding" the ground and life in the previously fertile Nile delta.

The oceans can also dissolve solids or suspend colloidal particles. However, the most important property of water, which presently prevents our immediate demise from air pollution, is the ocean's ability to absorb gases and other waste products. Table B gives the dissolved gases in the ocean water. It is important to keep in mind that the oceans can only absorb a certain finite quantity of carbon dioxide (CO_2) before they become saturated. The limited ability of the ocean to absorb CO_2 will affect us if we continue to burn hydrocarbon fuels. Once saturation occurs, then the tipping point has been reached, since the absorbed gases in the oceans will then be re-emitted into the atmosphere due to the ocean's increased temperature. This is the case because the solubility of gases in water are critically dependent on its temperature. Cooler water will absorb more gas.

The oceans now are studded with plastic and much insoluble content. They are also now carriers of radioactivity as a result of Fukushima.

The total content has increased substantially from what is given in Table A, which was compiled in 1976.

An increased amount of vapors in the atmosphere relates directly to the amount of sunlight being absorbed, via the "greenhouse" effect.

Table A.
The Major Dissolved Constituents of Sea Water

Ion	Percent
Cl	55.05
Na	30.61
SO_4	7.68
Mg	3.69
Ca	1.16
K	1.10
HCO3	0.41
Br	0.19
H_3BO_3	0.07
Sr	0.03
P	0.00
Total	**99.99**

Table B.
Dissolved Gases in Sea Water (1976 Data)
To Be Updated

	Concentration (cc/liter)
Oxygen	0 – 9
Nitrogen	8 – 15
Total carbon dioxide	34 – 56
Argon (residue after removal of N)	0.2 – 0.4
Helium and neon	1.7×10^{-4}
Hydrogen sulfide	variable

Once global oceanic temperatures rise, the solubility of CO_2 in the oceans decreases (slowly at first), resulting in an increase in the evaporation rate of ocean water. This, in turn, would increase the solar greenhouse effect, raising the temperature even more.

Once this "run-away" condition has set in, it will probably continue until all the glaciers and polar ice have melted (which would inundate all low-lying land masses - about 70% of the world's population). Then, with an increased water surface area (plus an increase in temperature), the evaporation of water will increase even more rapidly. With more vapor in the atmosphere, the atmospheric pressure would also increase.

First the facts. The measured CO_2 content in the atmosphere has been steadily increasing (Figures 14.1, 14.2). It seems to go up linearly, at first, as a result of fossil fuel burning. The steady increase indicates that our plant life cannot take up the increased CO_2 fast enough, nor are the oceans apparently capable of absorbing all of the excess—otherwise, we should see straight horizontal lines in Figures 14.1 and 14.2. Instead it goes up!! According to Fowler, CO_2 concentration was 290 ppm in the nineteenth century. It was projected to be 345 ppm by the year 2000—instead it rose to 370 ppm. It is now (2020) already above 400 ppm, We also have burned (ravaged) between 2 and 12% of our total fossil fuel reserves (there is uncertainty regarding the amount of reserves). Fowler projects that burning the rest would increase the CO_2 content in the atmosphere by as much as a factor of 4.

Examine the distribution of CO_2 in the biosphere. (Once all the oceans are evaporated, the atmospheric carbon [Table B] content would increase by a factor of 60, while the

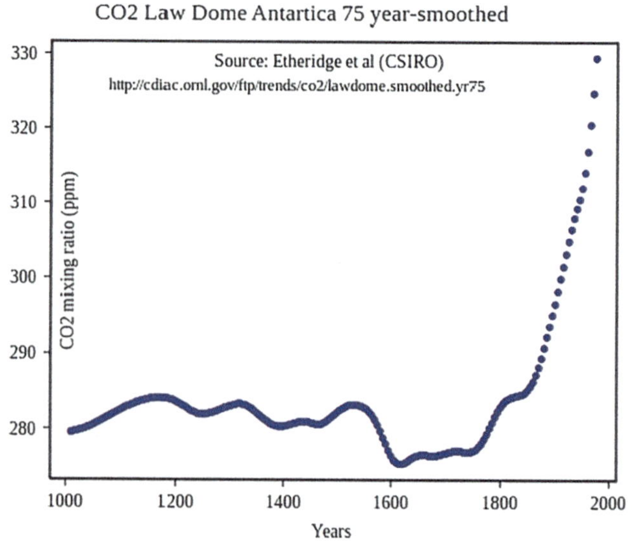

Figure 14.1

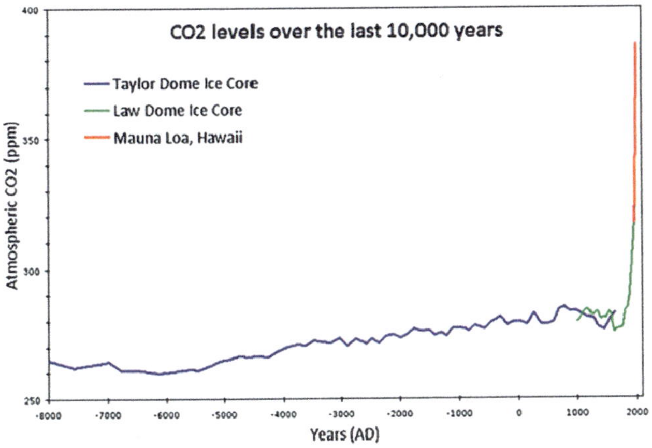

Figure 14.2

earth's atmosphere would initially be mainly water vapor, however.)

Note that roughly comparable amounts of carbon are distributed between the atmosphere, living matter, and the top mixing layer of the ocean.

Table C.
The Carbon Reservoir

Atmosphere	750 B tons
Oceans mixing layer	800 B tons
Living matter	1,740 B tons
Deep layers of ocean	43,500 B tons
Fossil fuel burning	4.4 B tons/year

(Fowler, J.W. 1975)

The argument often espoused is that CO_2 is good for plant life. That is true. However, an excess quantity of CO_2 certainly is not. Similarly, water too is essential for life, but too much water could drown us. Oxygen is essential but a 100% oxygen atmosphere is unsafe.

Some remarks regarding Figure 14.2 are relevant. Note the long-term increase of the CO_2 content over the last 6,000 years, which most likely was the result of advances in mankind's technology. Deforestation, building wooden structures, and burning wood all contributed to the rise in CO_2 on Earth. The Earth's population increase, in numbers and space utilization, also would have an effect. It is well-known that during the Roman period 2,000 years ago, a great amount of deforestation did take place, which could have contributed to the rise in CO_2 content. However, the Industrial Revolution and the enormous increase in the world's population during the last 100 years clearly made the largest impact. For roughly 100 years, the oceans were able to sustain CO_2 levels at a reasonable level.

The enormous increase in population and the use of fossil fuel for transportation purposes eventually overwhelmed the Earth's capacity to handle the excess CO_2 content. The tipping point for the major change occurred somewhere around 1980. From the ocean's point of view, the tipping point was about 1997 (See Figure 14.4).

It appears that the ocean temperature and its CO_2 content are the most critical elements to be considered.

Both the carbon and the hydrological water cycles are delicately coupled to the atmosphere: Minute changes in its content and temperature, could bring about major changes in our weather (Figure 14.3). This is called amplification. It appears that man's activities are now on such a sufficiently large scale that he can covertly or inadvertently bring about

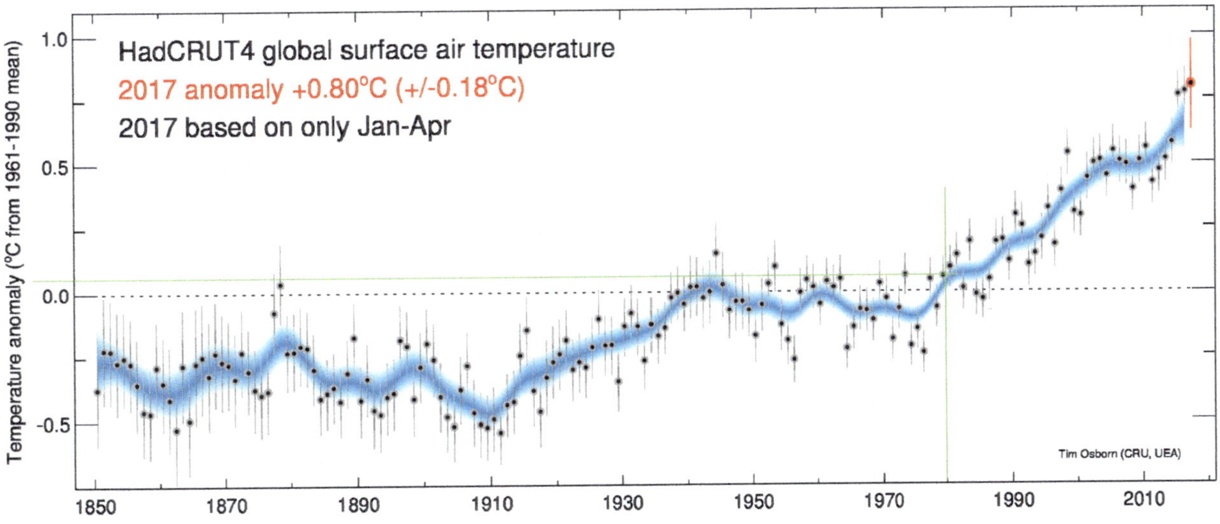

Figure 14.3

major changes. Unfortunately, meteorology may not be well enough understood or appreciated by the general public to provide the firm conviction needed.

There are already noticeable changes in the atmosphere's temperature (Figure 14.3).

Table D.
A Few of life's Key Constituents.

Mass of atmosphere	5×10^{21} gram
Mass of CO_2 in atmosphere	2×10^{18} gram
Mass of H_2O in atmosphere	Between $1-3\times10^{18}$ gram
Total fossil carbon (estimated)	7×10^{21} gram
Mass of O_2 in atmosphere	1×10^{21} gram
Recoverable carbon	$\sim8\times10^{18}$ gram
Solubility of CO_2 in water	
at 20° C	1 gram per 500 liter of H_2O
at 60° C	~1/4 gram per 500 liter of H_2O
Mass of ocean water (our reservoir)	1.5×10^{24} gram
Ocean area	3.6×10^{18} cm^2

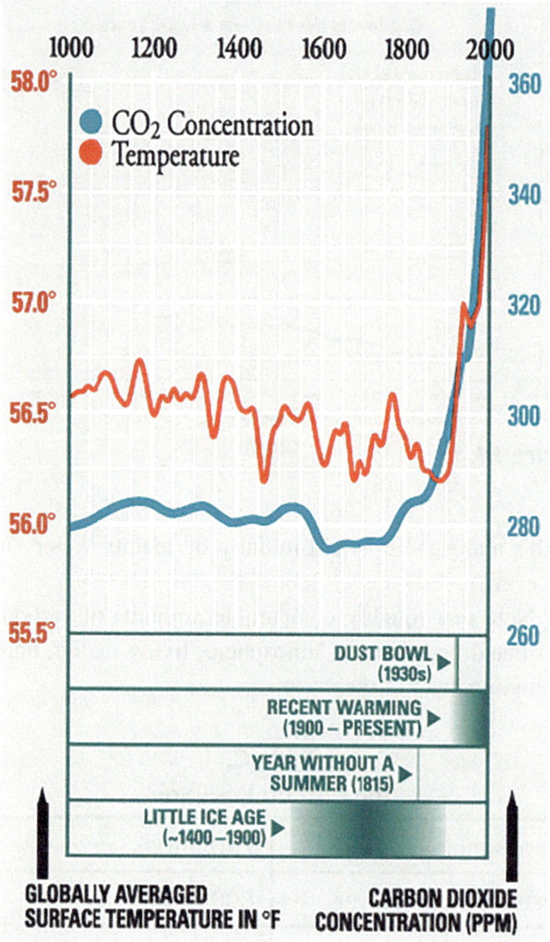

Figure 14.5

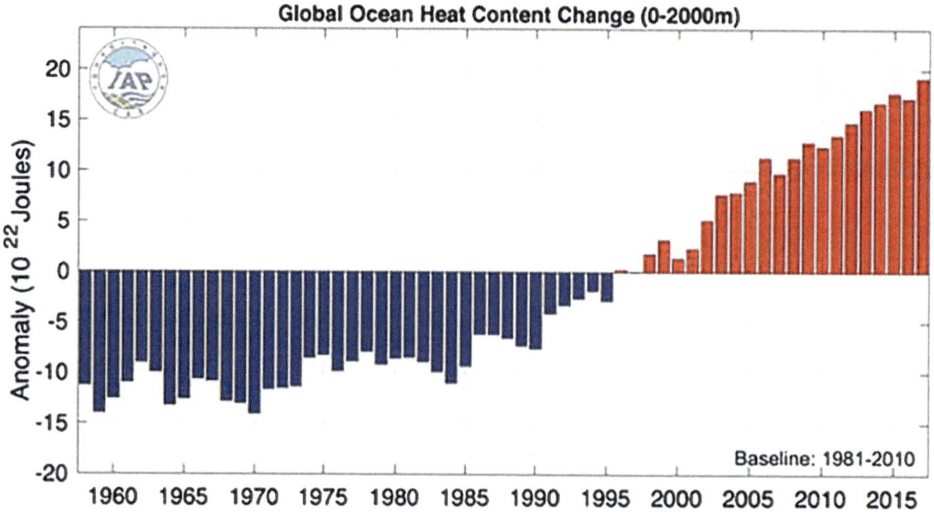

Figure 14.4

About 14 meter depth of ocean water has the mass of the whole atmosphere. An approximate 3 meter ocean depth is required to retain the circulating CO_2 supply in the ocean.

There was an increase of 0.6°C in the atmosphere temperature from the nineteenth century to 1940. Since then, it fluctuated by 0.2°C for a brief period in the northern hemisphere. Temperature changes may not seem spectacular, but remember, the energy balance and reactions in the atmosphere are so delicate that a mere 1 or 2% change in the energy available at the earth's surface can bring on a new ice age or melt the polar caps.

The atmosphere of the earth is clearly the "jugular" of the biosphere. We should obviously treat it with far greater respect and not tamper with it, unless its consequences are better understood. Study Figure 14.3. It looks rather ominous.

Next, I would like you to consider Figure 14.4, which displays the global ocean heat content. Note the incredibly large increase of its heat content, starting at about 1997, to the present. This increase inevitably would lead to atmospheric consequences by virtue of the fact that the oceans will increase their evaporation rate, along with the evaporation of some of its stored CO_2 content. This enormously large increase in heat content also implies a rise in temperature of the ocean's upper layers.

Have we reached the tipping point—the point of no return? Reexamine Fig.14.4 and note the change at about 1997. There is an increased slope without any retracement. This most likely suggests that unfortunately we may have passed the tipping point.

These conclusions are independent of any political philosophy. They are not intended to make a political statement, but instead represent an uncomplicated scientific analysis, based on the rate at which these changes are occurring

There are optimistic discussions afloat that we may be due for another ice age. However, it is unlikely that one will occur for another 100,000 years. Regardless, the coming of any such hopeful ice age, unfortunately, would most likely be too late in any event to avert the immediate, impending consequences.

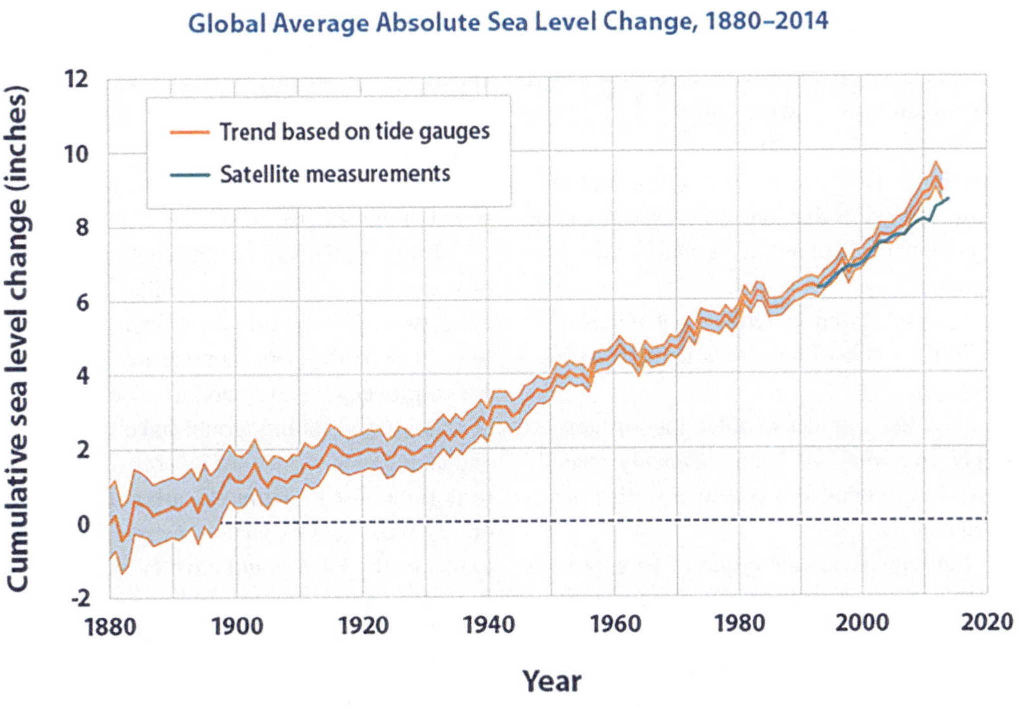

Data sources:

- CSIRO (Commonwealth Scientific and Industrial Research Organisation). 2015 update to data originally published in: Church, J.A., and N.J. White. 2011. Sea-level rise from the late 19th to the early 21st century. Surv. Geophys. 32:585–602. www.cmar.csiro.au/sealevel/sl_data_cmar.html.
- NOAA (National Oceanic and Atmospheric Administration). 2015. Laboratory for Satellite Altimetry: Sea level rise. Accessed June 2015. http://ibis.grdl.noaa.gov/SAT/SeaLevelRise/LSA_SLR_timeseries_global.php.

For more information, visit U.S. EPA's "Climate Change Indicators in the United States" at www.epa.gov/climatechange/indicators.

Figure 14.6 *Sea Level Rise*

Venus

Credit: NASA, The atmosphere of Venus as viewed in ultraviolet by the Pioneer Venus Orbiter in 1979.

We shall now examine the fate of our sister planet Venus, which could have had a similar scenario. However, its causation was most likely different.

A rough calculation of the total amount of carbon associated with the carbon dioxide content of Venus' atmosphere is roughly comparable in magnitude with the total amount of available carbon associated with the Earth (Table C). This is a rather ominous fact.

Remember, this is our neighbor, our sister planet. It developed for billions of years just like the Earth, with mountains, valleys and probably also oceans. Radar photography confirmed a complicated terrain on Venus. However, here are differences between Venus and the Earth:

a. Venus does not have a measurable magnetic field.
b. It almost stopped its rotation completely. A day on Venus comprises 243 Earth days, and in the opposite direction.
c. It clearly displays an incredibly hot atmosphere (about 700° F), consisting of 98% carbon dioxide.

The above are our clues in order to solve this mystery. A severe catastrophe overwhelmed Venus relatively recently since its geological development was similar to that of the Earth's. What happened?

In order to fully appreciate the origin of an enormous disaster on Venus, one has to study the functioning of planetary cores in general. Here our Earth, as usual, provides important clues: Each core spins independently from its outer mantle/crust, being insulated by a high-pressure gaseous region (on Earth, this region is designated as the "outer core"). Each core acts like a giant gyroscope with an electric outer charge, which, by virtue of its rotation, produces the well-known planetary magnetic fields. Cores have an opportunity to flip 180° many times during their 4.5 billion year development, based on measured magnetic field reversals recorded adjacent to the edges of ocean trenches. Core reversals require some external torque mechanism (perhaps in the form of a release of matter or a bulge emanating from the solid core, which extends through the gaseous outer core to momentarily impact the Earth's solid mantle), resulting in a change in spin orientation of the inner core.

Assume that for Venus, its core was spinning in an opposite direction from its outer mantle, and further assume that this solid core made sudden permanent contact with its outer mantle. This would almost cancel its total angular momentum, where its final direction of rotation would be governed by the net total angular momentum remaining. All this, of course, is speculation. The event described above would have secondary consequences: it would produce an enormous amount of heat, which it would impart to the surrounding gas, which in turn will seek a way out through any of the existing internal fissures. When this extremely hot hydrocarbon gas encounters Venus's oxygen atmosphere, an enormous conflagration must have ensued, which would burn all available carbon present on Venus's surface, in addition to the hydrocarbons which emanated from the planet's core. The surface of Venus must have been exceedingly hot, which would also have evaporated its potential oceans. Venus's water vapor in the upper atmosphere would become dissociated via the Sun's ultraviolet radiation into hydrogen and oxygen molecules, with the hydrogen gas escaping from the planet. However, the heavier rare gas elements display abundances on both the Earth and Venus, which are remarkably similar in their comparable magnitudes (Cox 2000 p. 301, Table 12.10)

There is another possible (but far less likely) variation of Venus's disaster scenario: Unlike the Earth, Venus did not avail itself of the opportunity to launch a satellite subsequent to its birth (perhaps its spin rate was too small to allow it to jettison its equatorial material). The eventual consequence of this alleged failure would have been an increased heat and energy content within its core, which could have been responsible for its ultimate carbon dioxide catastrophe. In other words, Venus's greater internal energy content relative to that of the Earth could have eventually produced a large enough internal pressure to release its gases through the surrounding mantle, possibly even influencing the planet's core spin during this release, via some unspecified mechanism.

The question of possible past "life" on Venus might be of interest. One may assume that Venus had a similar development to that of the Earth: Oceans studded with potential organic content, could have evolved into living organisms, given sufficient time. Since Venus's disaster happened in recent eons, one might even assume that life could have developed, but is presently and permanently extinguished due to the surge of heat and carbon dioxide in its atmosphere. Of course there is always the remote possibility that some extremely hardy life might have survived, but most likely not.

References

1. Cox, A.N. 2000. Allen's Astrophysical Quantities. Fourth Edition. Springer-Verlag New York, Inc. New York.

2. Fowler, J.M. 1975, *Energy and the Environment*, McGraw Hill.

Chapter 15

Conclusions

Synopsis
Failures
Things to Ponder About the Universe

This book covers a very carefully selected set of topics, meant to develop an **internally self-consistent** theme. The theme is evolution, from the neutron cores produced at the beginning of the Cosmos, to the arrival of mankind.

This represented a significant departure from the present paradigm, and therefore required a different perspective, a **forensic** approach, and a more critical analysis of the available data. Many of the most popular astronomical theories were shown to be incorrect. Such as the Big Bang, Dark Energy, Black Holes, Dark Matter and Accretion Discs.

The all-encompassing perspective centered around the idea that there is undeniably more involved in the evolution of mankind than simply chance.

This text should not be the end of the discussion. There is plenty of work left to do. I strongly encourage others to carry on with the torch that I have lit.

It has been a long time since a major shift in our conception of the universe has undergone a change of major proportions. Just when the comfort level of astronomers is such that most of the them would routinely recite their *beliefs* of (imagined) cosmic evolution, without even a hint of a doubt: "Dust clouds and their collapse led to the origin of solar systems" is, at the present time, the accepted evolutionary scenario.

This author is a reluctant spoiler. Unfortunately, however, even skeptics must realize that some laws of physics cannot be readily ignored nor circumvented. These include the Second Law of Thermodynamics and the Conservation of Angular Momentum of an isolated system. Thus, the present distribution of angular momentum in our solar system is such that given these two constraints, the solar system could not have evolved from a collapsing cloud. It is a seriously flawed theory. Moreover, if this is the case, and, if our solar system is considered to be typical of other planetary systems, it is difficult to escape the conclusion that most interstellar clouds are not the nurseries for stellar evolution, as is generally believed, but the dumping grounds or " leftovers" of prior stellar developments. Perhaps the greatest fallacy is the assumption that our Sun and the planets of our solar system were devoid of sophisticated development since their origin.

Failures

There are numerous examples of failed developments at every evolutionary stage, all over the universe.

The Earth experienced many such disastrous events during its development, evidence of which is well-recorded. The Earth's biological history includes many examples of species that have become extinct (trilobites and dinosaurs are typical examples).

The Asteroid Belt is another example. There could have been a planet in orbit that failed its development. Venus is an example of a planet whose initial development seemed similar to that of the Earth. However, Venus underwent a prior carbon dioxide disaster, which would have eliminated any existing life and stymied the possibility of life's development (at least as we currently understand it). I am not quite sure whether our planet Earth will be among the failures.

As for stellar developments in general, we know that within our own galaxy there are at least 600 pulsars, indicative of failed stellar developments, since each pulsar is the result of a supernova explosion. Astronomical studies of deep space objects encounter some highly dramatic events within galactic developments. It is therefore not surprising to see that an incredible redundancy factor is built into each developmental stage, knowing that failure is probable with every evolutionary step.

> This could provide a possible insight into a higher power that we may call the Creating Entity of the Universe (CE). The overall plan, it seems, is to have a multitude of possibilities in each development stage - in the expectation that some of them might be successful. This has further implications. Namely, if one intends to launch such an

enormously complex "Mega system", would one not like to see results? Hence, there is the remote possibility that a monitoring system might be in existence by which CE would monitor the progress of on-going experiments. If this were the case, it would be entirely outside our very limited knowledge of physics.

Based on our present knowledge of physics, and observations of our vast universe, consider the following:

- The laws of physics are applicable throughout the universe.
- Laws of chemistry are based on physics. We can recognize molecules throughout the universe. Hence, spectroscopy based on quantum mechanics is also universal.
- The age of the Earth is 4 ½ billion years. It took a long time for Homo sapiens to evolve and reach this age of technology (which is probably just the tip of the iceberg).
- According to the author's *Similarity Principle*, the duplication of a multitude of solar system developments is to be expected. However, its details and the timing of each one of them would differ substantially.

Things to Ponder About the Universe

1. Who in the cosmos has the capability to carry out and launch this enormous, complex, project, assuming that its development has a purpose?
2. What would be the purpose for doing this?
3. Can one draw some inference based on our limited knowledge?
4. What is the basic material from which everything was constructed?

Based on a biological analogy, it seems that the most fundamental structure could be the neutron, which contains a proton, electron, neutrino, and other possible information carriers, all analogous to the DNA (or an egg) in a biological organism. Thus, somewhere in the universe, a production site for these neutrons would exist for their assembly, to form highly massive and dense objects (this being the "winding-up process"). These dense objects disintegrate or unwind, which in turn provides the energy and material necessary for the formation of the universe.

This is about as far as we can go at this time, since we have no idea where or how neutrons were created, the ephemeral location of a production site, or where the matter that composed them may have originated. These questions have no answers.

In view of the multiplicity of venues and uncertainty in the outcome of each individual galaxy and each of its diverse solar systems, a redundancy factor makes sense. Of course, the passage of time would not be operating for a Super Entity (time however, is severely limited to us mortals). Could CE be an experimentalist (or maybe a comedian)?

The physical location of our Super Entity is a mystery and completely unknown to any of us mortals. Could it be dispersed among some of its multiple components?

Another mystery: Where did all the basic matter which constitutes our universe come from, including the Super Entity itself? These questions were being raised multiple times in the past and are still being considered to-day, without a meaningful response. However, this may be the underlying popularity for the "instant poof" concept, as an extreme option for both faith based Biblical and Big Bang creation scenarios.

What seems to me the most remarkable of all, is that we are but a collection of molecules that have evolved during 4.5 billion years and finally developed the ability to investigate (at least to some extent) our own origin and to communicate this to others. This in itself is a remarkable accomplishment.

Please also consult my conclusions in the astronomy textbook "Voyage into Astronomy" (2015).

Perhaps, this text will also provide a better appreciation of our own lives and our small place in this vast undertaking. There is always the possibility that there might be a purpose behind this entire complicated development scenario and that reverence for that almighty establisher of the universe is not misplaced.

If our universe was established by a Super Entity (also known as God), it was wound up in a tight-knit, rapidly spinning "neutron ball", then launched. Its vast components are presently in multiple developing stages and simultaneously winding down—just as this text must finally do.

Thank you for having come along with me on this adventure: a forensic journey of discovery and enlightenment.

Chapter 16

Definitions

The two important laws in Thermodynamics are the following:

- **The First Law:** states that all forces have to be accounted for. There is no "free lunch."
- **The Second Law:** however is more challenging. It involves the concept of ENTROPY which essentially refers to the fact that everything in the universe (without exception) becomes more disorganized, never the reverse. "All natural (i.e. irreversible) processes take place with an increase in entropy" (University Physics 6th Ed. Sears, Zemansky and Young, P. 380). (The concept of ENTROPY clearly plays a central role in all of Astronomy, as will be shown in this text.
- **Angular Momentum:** (L) is defined as the product of mass (m), radius (r) and velocity (v). In a closed system, the quantity L remains constant. **L=m•v•r**
- **Fission:** The spitting of a heavy unstable nuclei (e.g.$_{92}$ **U**^{235}) into lighter atoms, accompanied by the release of energy.
- **Fusion:** Is the process where elementary particles such as protons, deuterium or tritium are forced to merge into heavier atoms under the influence of both high pressure and 15 million degree temperatures.
- **The neutrino:** Is a neutral (no charge) subatomic particle with almost zero mass and spin of 1/2, which very rarely interacts with normal matter.
- **Photon:** Is a particle representing a quantum of electromagnetic radiation. It has zero rest mass and energy proportional to its frequency.
- **Luminosity:** Is the total amount of electromagnetic energy emitted per second by any celestial object such as a planet, a star or a galactic center.
- **The Doppler:** effect is explained in Chapter 9
- **The Hubble Relationship:** is also explained in Chapter 9
- **The Ventury Effect** is the reduction in gas or fluid pressure as it exists a constricted opening: High pressure gas can thereby be converted to liquid.
- **The Stefan-Bolzmann Law:** states that the total radiation by a black body is proportional to the fourth power of its absolute temperature: $E=\sigma T^4$ per unit area, where $\sigma=5.67\times 10^{-8}$ W m^{-2} K^{-4}.
- **The Hydrogen Balmer lines:** in atomic physics, are one of a set of six named series describing the wavelengths of spectral line emissions from an exited hydrogen atom. The Balmer series is calculated using the Balmer formula, an empirical equation discovered by Johann Balmer in 1885.

Index

Italic page numbers indicate images.

G

H

I

J

K

L

N

O

T

U